**IEE PROFESSIONAL APPLICATIONS
OF COMPUTING SERIES 5**

Series Editors: Professor P. Thomas
Dr R. Macredie
J. Smith

Intelligent Distributed Video Surveillance Systems

Other volumes in this series:

Intelligent Distributed Video Surveillance Systems

Edited by
Dr Sergio A. Velastin and
Dr Paolo Remagnino

The Institution of Electrical Engineers

Published by: The Institution of Electrical Engineers, London,
United Kingdom

The Institution of Electrical Engineers,
Michael Faraday House,
Six Hills Way, Stevenage,
Herts., SG1 2AY, United Kingdom

www.iee.org

British Library Cataloguing in Publication Data

Intelligent distributed video surveillance systems. – (IEE computing series; PC005)
   1. Computer Vision
   I. Velastin, Sergio II. Remagnino, Paolo, 1963–
   III. Institution of Electrical Engineers
   006.3'7

ISBN-10: 0 86341 504 0
ISBN-13: 978-086341-504-3

Typeset in India by Newgen Imaging Systems (P) Ltd., Chennai, India
Printed in the UK by MPG Books Ltd., Bodmin, Cornwall

Contents

7   A distributed multi-sensor surveillance system for public transport applications   185

J-L. Bruyelle, L. Khoudour, D. Aubert, T. Leclercq and A. Flancquart

Foreword

S. A. Velastin and P. Remagnino

Conventional video surveillance, where people are asked to sit in front of batteries of TV monitors for many hours, will soon become obsolete. This is a reasonable assumption, given the recent advances in machine intelligence research, the very fast technology progress and the increasing deployment rate of surveillance cameras and networks throughout Europe and indeed worldwide.

Two particular aspects of machine intelligence have received attention and improved considerably over the last decade. On the one hand the development of machine vision algorithms capable of handling complex visual data, acquired by camera systems, and delivering human-like commentaries of the evolution of the monitored scene; on the other hand the advances in distributed computing and distributed intelligence, capable of handling devices as separate entities, capable of adapting to the evolution of the scene and the complexity of the communication network, reducing information bandwidth and inferring a better interpretation of the dynamics of people and objects moving in the scene. The latter, in particular, is an important prerequisite for addressing the inherently distributed nature of the surveillance process. Take for example the monitoring of public spaces such as a city centre or a metropolitan railway network. One major urban station can typically have 100 cameras and two control rooms where a person is typically required to monitor 20 to 40 cameras at any given time plus carrying out other duties such as public announcements, keeping traffic control informed of problems, filling up reports and coordinating the work of ground staff. Personnel dealing with specific areas such as platforms might also have access to a reduced number of cameras and other information sources (such as the destination of the next three trains) and deal with local problems by themselves unless they merit the attention of the station control room, who in turn manage problems with their own resources unless they need to involve other personnel such as a station group manager, a line manager or the police. The lessons to be learned for those currently attempting to automate the monitoring task is that surveillance systems have an inherent distributed nature (geographical and managerial), decisions on how to intervene being made by groups of people in hierarchical and parallel organisations are based on a diverse set of data, not only visual but also through voice, textual and contextual.

It is clear that major progress has been made on computer vision techniques for the analysis of human activity and biometrics, as witnessed by the large number of papers written on these subjects and high-impact technologies such as automatic number plate recognition (APNR) and face recognition systems. However, practical visual surveillance solutions need to be integrated as part of potentially very large systems in ways that are scalable, reliable and usable to the people operating such systems. Technically, this involves making important bridges between the computer vision, networking and system engineering communities. Finally, it is also important that the scientific and technical developments are fully aware of aspects related to operational practice, legal considerations, ethical issues and the aspirations and expectations of the public at large.

With this in mind, two symposia were organised by the Institution of Electrical Engineers (IEE) in 2003 and 2004 and called Intelligent Distributed Surveillance Systems (IDSS). The meetings dealt with the latest developments in distributed intelligent surveillance systems, in terms of technical advances, practical deployment issues and wider aspects such as implications for personal security and privacy. They brought together presentations from researchers and engineers working in the field, system integrators and managers of public and private organisations likely to use such systems.

This book then brings together, in what we hope is a self-contained collection of the latest current work in this field, extended versions of papers presented at the IDSS. We expect it will appeal equally to computer vision researchers, scientists and engineers as well as those concerned with the deployment of advanced distributed surveillance systems.

The book is organised in nine chapters, each describing a method, an implementation or a study of a distributed system. All chapters discuss algorithms and implementations employed not only in classic video surveillance but also in the new emerging research field of ambient intelligence.

In Chapter 1 Valera and Velastin set the scene by reviewing the state-of-the-art in automatic video surveillance systems and describe what new research areas and disciplines will have to take part in the design of future and wider-scale distributed surveillance systems.

Svensson, Heath and Luff, in Chapter 2, study the importance and social and psychological implications of modern video surveillance systems. Their limitations are discussed in particular for conventional control rooms, together with public awareness of the video surveillance systems, now becoming widespread.

Black, Ellis and Makris, in Chapter 3, introduce the first technical issue studied in this collection: the importance of a flexible infrastructure, to move and process efficiently video data across a wide area. Such infrastructure has to operate in real-time and offer high-capacity storage to save raw video data and related processing.

In Chapter 4 Cucchiara, Grana, Prati and Vezzani discuss a particular implementation of the emerging ambient intelligence (AmI) paradigm. AmI supports the design of a smart environment to aid the users inhabiting the augmented environment. This chapter covers the area of domotics, from *domus* (home) and *informatics*. The shift of interest is from security to safety and the well being of the user.

Desurmont, Bastide, Czyz, Parisot, Delaigle and Macq discuss the problem of data communication in Chapter 5. This chapter complements Chapter 3 by offering solutions to data communication in large video surveillance systems.

Chapter 6, by Bowden, Gilbert and KaewTraKulPong, focuses on the importance of the topology of the acquiring sensor array and covers a study on the automatic calibration of such networks, with more or less overlapping fields of view.

In Chapter 7, Bruyelle, Khoudour, Aubert, Leclercq and Flancquart introduce and discuss the deployment of a large surveillance system for public transport applications. The chapter reports a study carried out in the European-funded project PRISMATICA.

Chapter 8, by Thirde, Xu and Orwell, offers another example of ambient intelligence. In this chapter, monitoring techniques are applied to football. A network of cameras, installed in a stadium, acquires video data of football matches, and conventional and adapted video surveillance techniques are employed to provide a real-time description of the match evolution.

In Chapter 9 Foresti, Micheloni, Snidaro and Piciarelli describe a hierarchical sensor array infrastructure, capable of controlling both fixed and mobile cameras. The infrastructure is organised as an ensemble of nodal points, corresponding to clusters of sensors and the cooperation between nodal points in discussed.

We hope that you enjoy the book as much as we have enjoyed reading and editing all the contributions. We are grateful to all the authors, to the IEE's Visual Information Engineering Professional Network (its executive committee, professional network manager and members) for supporting IDSS-03 and IDSS-04, to Michel Renard of VIGITEC for their support to Sarah Kramer of the IEE Publishing team for her initiative and assistance and to the IEE's Visual Information Engineering Professional Network Executive for their support.

September 2005
Digital Imaging Research Centre
Kingston University, UK

Authors and Contributors

M. Valera
Digital Imaging Research Centre
School of Computing and Information
Systems
Kingston University, UK

J. Black
Digital Imaging Research Centre
Kingston University
Kingston-upon-Thames
Surrey, UK

S.A. Velastin
Digital Imaging Research Centre
School of Computing and Information
Systems
Kingston University, UK

T. Ellis
Digital Imaging Research Centre
Kingston University
Kingston-upon-Thames
Surrey, UK

M. S. Svensson
Work, Interaction and Technology
Research Group,
The Management Centre
King's College London, UK

D. Makris
Digital Imaging Research Centre
Kingston University
Kingston-upon-Thames
Surrey, UK

C. Heath
Work, Interaction and Technology
Research Group,
The Management Centre
King's College London, UK

R. Cucchiara
Dipartimento di Ingegneria
dell'Informazione
Università di Modena e Reggio Emilia
Modena, Italy

P. Luff
Work, Interaction and Technology
Research Group,
The Management Centre
King's College London, UK

C. Grana
Dipartimento di Ingegneria
dell'Informazione
Università di Modena e Reggio Emilia
Modena, Italy

A. Prati
Dipartimento di Ingegneria
dell'Informazione
Università di Modena e Reggio Emilia
Modena, Italy

R. Vezzani
Dipartimento di Ingegneria
dell'Informazione
Università di Modena e Reggio Emilia
Modena, Italy

X. Desurmont
Multitel, Parc Initialis – Avenue
Copernic, Mons, Belgium
Universite Catholique de Louvain,
Belgium

A. Bastide
Multitel, Parc Initialis – Avenue
Copernic, Mons, Belgium
Universite Catholique de Louvain,
Belgium

J. Czyz
Multitel, Parc Initialis – Avenue
Copernic, Mons, Belgium
Universite Catholique de Louvain,
Belgium

B. Macq
Multitel, Parc Initialis – Avenue
Copernic, Mons, Belgium
Universite Catholique de Louvain,
Belgium

C. Parisot
Multitel, Parc Initialis – Avenue
Copernic, Mons, Belgium
Universite Catholique de Louvain,
Belgium

J-F. Delaigle
Multitel, Parc Initialis – Avenue
Copernic, Mons, Belgium
Universite Catholique de Louvain,
Belgium

R. Bowden
CVSSP,University of Surrey Guildford,
Surrey, UK

A. Gilbert
CVSSP,University of Surrey Guildford,
Surrey, UK

P. KaewTraKulPong
King Mongkut's University of
Technology Thonburi,
Faculty of Engineering,Toongkaru,
Bangmod,
Bangkok, Thailand

J-L. Bruyelle
INRETS-LEOST, 20 rue Elisée Reclus,
Villeneuve d'Ascq, France

L. Khoudour
INRETS-LEOST, 20 rue Elisée Reclus,
Villeneuve d'Ascq, France

T. Leclercq
INRETS-LEOST, 20 rue Elisée Reclus,
Villeneuve d'Ascq, France

A. Flancquart
INRETS-LEOST, 20 rue Elisée Reclus,
Villeneuve d'Ascq, France

D. Aubert
INRETS-LIVIC, route de la Minière,
Versailles, France

D. Thirde
Digital Imaging Research Centre
Kingston University
Kingston upon Thames, UK

M. Xu
Digital Imaging Research Centre
Kingston University
Kingston upon Thames, UK

J. Orwell
Digital Imaging Research Centre
Kingston University
Kingston upon Thames, UK

G.L. Foresti
Department of Computer Science,
University of Udine, Italy

C. Micheloni
Department of Computer Science,
University of Udine, Italy

L. Snidaro
Department of Computer Science,
University of Udine, Italy

C. Piciarelli
Department of Computer Science,
University of Udine, Italy

Chapter 1

A review of the state-of-the-art in distributed surveillance systems

M. Valera and S. A. Velastin

1.1 Introduction

Intelligent visual surveillance systems deal with the real-time monitoring of persistent and transient objects within a specific environment. The primary aims of these systems are to provide an automatic interpretation of scenes and to understand and predict the actions and interactions of the observed objects based on the information acquired by sensors. The main stages of processing in an intelligent visual surveillance system are moving object detection and recognition, tracking, behavioural analysis and retrieval. These stages involve the topics of machine vision, pattern analysis, artificial intelligence and data management.

The recent interest for surveillance in public, military and commercial scenarios is increasing the need to create and deploy intelligent or automated visual surveillance systems. In scenarios such as public transport, these systems can help monitor and store situations of interest involving the public, viewed both as individuals and as crowds. Current research in these automated visual surveillance systems tends to combine multiple disciplines such as those mentioned earlier with signal processing, telecommunications, management and socio-ethical studies. Nevertheless there tends to be a lack of contributions from the field of system engineering to the research.

The growing research interest in this field is exemplified by the IEEE and IEE workshops and conferences on visual surveillance [1–6] and special issues that focus solely on visual surveillance in journals like [7–9] or in human motion analysis like in [10]. This chapter surveys the work on automated surveillance systems from the aspects of

- image processing/computer vision algorithms which are currently used for visual surveillance;

- surveillance systems: different approaches to the integration of the different vision algorithms to build a completed surveillance system;
- distribution, communication and system design: discussion of how such methods need to be integrated into large systems to mirror the needs of practical CCTV installations in the future.

Even though the main goal of this chapter is to present a review of the work that has been done in surveillance systems, an outline of different image processing techniques, which constitute the low-level part of these systems, is included to provide a better context. One criterion of classification of surveillance systems at the sensor level (signal processing) is related to sensor modality (e.g. infrared, audio and video), sensor multiplicity (stereo or monocular) and sensor placement (centralised or distributed). This review focuses on automated video surveillance systems based on one or more stereo or monocular cameras because there is not much work reported on the integration of different types of sensors such as video and audio. However some systems such as [11,12] process the information that comes from different kinds of sensor such as audio and video. After a presentation of the historical evolution of automated surveillance systems, Section 1.2 identifies different typical application scenarios. An outline of different image processing techniques that are used in surveillance systems is given in Section 1.3. A review of surveillance systems, which this chapter focuses on, is then presented in Section 1.4. Section 1.5 provides a discussion of the main technical issues in these automated surveillance systems (distribution and system design aspects) and presents approaches taken in other fields like robotics. Finally, Section 1.6 concludes with possible future research directions.

The technological evolution of video-based surveillance systems started with analogue CCTV systems. These systems consist of a number of cameras located in a multiple remote location and connected to a set of monitors, usually placed in a single control room, via switches (a video matrix). In Reference 13, for example, integration of different CCTV systems to monitor transport systems is discussed. Currently, the majority of CCTV systems use analogue techniques for image distribution and storage. Conventional CCTV cameras generally use a digital charge coupled device (CCD) to capture images. The digital image is then converted into an analogue composite video signal, which is connected to the CCTV matrix, monitors and recording equipment, generally via coaxial cables. The digital-to-analogue conversion does cause some picture degradation, and the analogue signal is susceptible to noise. It is possible to have CCTV digital systems by taking advantage of the initial digital format of the captured images and by using high-performance computers. The technological improvement provided by these systems has led to the development of semi-automatic systems, known as second generation surveillance systems. Most of the research in second generation surveillance systems is based on the creation of algorithms for automatic real-time detection events aiding the user to recognise the events. Table 1.1 summarises the technological evolution of intelligent surveillance systems (first, second and third generation), outlining the main problems and the current research in each of them.

Table 1.1   *Summary of the technical evolution of intelligent surveillance systems*

First generation	
Techniques	Analogue CCTV systems
Advantages	Give good performance in some situations
	Mature technology
Problems	Use analogue techniques for image distribution and storage
Current research	Digital versus analogue
	Digital video recording
	CCTV video compression
Second generation	
Techniques	Automated visual surveillance by combining computer vision technology with CCTV systems
Advantages	Increase the surveillance efficiency of CCTV systems
Problems	Robust detection and tracking algorithms required for behavioural analysis
Current research	Real-time robust computer vision algorithms
	Automatic learning of scene variability and patterns of behaviours
	Bridging the gap between the statistical analysis of a scene and producing natural language interpretations
Third generation	
Techniques	Automated wide-area surveillance system
Advantages	More accurate information as a result of combining different kinds of sensors
	Distribution
Problems	Distribution of information (integration and communication)
	Design methodology
	Moving platforms
	Multi-sensor platforms
Current research	Distributed versus centralised intelligence
	Data fusion
	Probabilistic reasoning framework
	Multi-camera surveillance techniques

1.2   Applications

The increasing demand for security from society leads to a growing need for surveillance activities in many environments. Recently, the demand for remote monitoring for safety and security purposes has received particular attention, especially in the following areas:

- Transport applications such as airports [14,15], maritime environments [16,17], railways, underground [12,13,18–20], and motorways to survey traffic [21–25].

- Public places such as banks, supermarkets, department stores [26–31] and parking lots [32–34].
- Remote surveillance of human activities such as attendance at football matches [35] or other activities [36–38].
- Surveillance to obtain certain quality control in many industrial processes, surveillance in forensic applications [39] and remote surveillance in military applications.

Recent events, including major terrorist attacks, have led to the increase in demand for security in society. This in turn has forced governments to make personal and asset security a priority in their policies. This has resulted in the deployment of large CCTV systems. For example, London Underground and Heathrow Airport have more than 5000 cameras each. To handle this large amount of information, issues such as scalability and usability (how information needs to be given to the right people at the right time) become very important. To cope with this growing demand, research and development has been continuously carried out in commercial and academic environments to find improvements or new solutions in signal processing, communications, system engineering and computer vision. Surveillance systems created for commercial purposes [26,28] differ from surveillance systems created in the academic world (e.g. [12,18,40,41]) where commercial systems tend to use specific-purpose hardware and an increasing use of networks of digital intelligent cameras. The common processing tasks that these systems perform are intrusion and motion detection [11,42–46] and detection of packages [42,45,46]. A technical review of commercial surveillance systems for railway applications can be found in Reference 47.

Research in academia tends to focus on the improvement of image processing tasks by making algorithms more accurate and robust in object detection and recognition [34–52], tracking [34,38,48,53–56], human activity recognition [57–59] and database [60–62] and tracking performance evaluation tools [63]. In Reference 64 a review of human body and movement detection, tracking and human activity recognition is presented. Other research currently carried out is based on the study of new solutions for video communication in distributed surveillance systems. Examples of these systems are video compression techniques [65–67], network and protocol techniques [68–70], distribution of processing tasks [71] and possible standards for data formats to be sent across the network [12,18,62]. The creation of a distributed automatic surveillance system by developing multi-camera or multi-sensor surveillance systems, and fusion of information obtained across cameras [12,36,41,72–76], or by creating an integrated system [12,19,53] is also an active area of research.

1.3 Techniques used in surveillance systems

This section summarises research that addresses the main image processing tasks that were identified in Section 1.2. A typical configuration of processing modules is

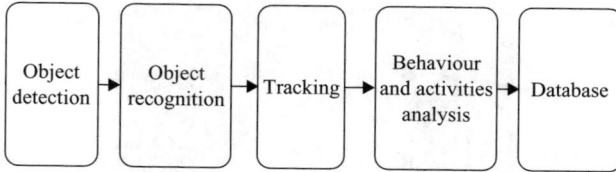

Figure 1.1   Traditional flow of processing in a visual surveillance system

Figure 1.2   Example of a temporal difference technique used in motion detection

illustrated in Figure 1.1. These modules constitute the low-level building blocks necessary for any distributed surveillance system. Therefore, each of the following sections outlines the most popular image processing techniques used in each of these modules. The interested reader can consult the references provided in this chapter for more details on these techniques.

1.3.1   Object detection

There are two main conventional approaches to object detection: 'temporal difference' and 'background subtraction'. The first approach consists in the subtraction of two consecutive frames followed by thresholding. The second technique is based on the subtraction of a background or reference model and the current image followed by a labelling process. After applying one of these approaches, morphological operations are typically applied to reduce the noise of the image difference. The temporal difference technique has good performance in dynamic environments because it is very adaptive, but it has a poor performance in extracting all the relevant object pixels. On the other hand, the background subtraction has a better performance in extracting object information but it is sensitive to dynamic changes in the environment (see Figures 1.2 and 1.3).

An adaptive background subtraction technique involves creating a background model and continuously upgrading it to avoid poor detection when there are changes in the environment. There are different techniques for modelling the background, which are directly related to the application. For example, in indoor environments with good lighting conditions and stationary cameras, it is possible to create a simple

Figure 1.3   Example of a background subtraction technique used in motion detection. In this example a bounding box is drawn to fit the object detected

background model by temporally smoothing the sequence of acquired images in a short time as described in [38,73,74].

Outdoor environments usually have high variability in scene conditions, and thus it is necessary to have robust adaptive background models, even though these robust models are computationally more expensive. A typical example is the use of a Gaussian model (GM) that models the intensity of each pixel with a single Gaussian distribution [77] or with more than one Gaussian distribution (Gaussian mixture models). In Reference 34, due to the particular characteristics of the environment (a forest), they use a combination of two Gaussian mixture models to cope with a bimodal background (e.g. movement of trees in the wind). The authors in Reference 59 use a mixture of Gaussians to model each pixel. The method they adopted handles slow lighting changes by slowly adapting the values of the Gaussians. A similar method is used in Reference 78. In Reference 54 the background model is based on estimating the noise of each pixel in a sequence of background images. From the estimated noise the pixels that represent moving regions are detected.

Other techniques of object detection use groups of pixels as the basic units for tracking, and the pixels are grouped by clustering techniques combining colour information (R,G,B) and spatial dimension (x, y) to make the clustering more robust. Algorithms such as expectation minimisation (EM) are used to track moving objects as clusters of pixels significantly different from the corresponding image reference, for example, in Reference 79 the authors use EM to simultaneously cluster trajectories belonging to one motion behaviour and then to learn the characteristic motions of this behaviour.

In Reference 80 the reported object detection technique is based on wavelet coefficients for detecting frontal and rear views of pedestrians. By using a variant of Haar wavelet coefficients as a low-level process of the intensity of the images, it is possible to extract high-level information on the object (pedestrian) to detect, for example, shape information. In a training stage, the coefficients that most accurately represent the object to be detected are selected using large training sets. Once the best coefficients have been selected, they use a support vector machine (SVM) to classify the training set. During the detection stage, the selected features are extracted from the image and then the SVM is used to verify the detection of the object. The advantage of using wavelet techniques is that of not having to rely on explicit colour information or textures. Therefore they can be useful in applications where there is a lack of colour information (a usual occurrence in indoor surveillance). Moreover, using wavelets implies a significant reduction of data in the learning stage. However, the authors only model the front and the rear views of pedestrians. In the case of groups of people that stop, talk or walk perpendicular to the view of the camera, the algorithm is not able to detect the people. Furthermore, an object with intensity characteristics similar to those of a frontal or rear human is likely to generate a false positive. Another line of research is based on the detection of contours of persons by using principal component analysis (PCA). Finally, as far as motion segmentation is concerned, techniques based on optic flow may be useful when a system uses moving cameras as in Reference 25, although there are known problems when the image size of the objects to be tracked is small.

1.3.2 Object recognition, tracking and performance evaluation

Tracking techniques can be split in two main approaches: 2D models with or without explicit shape models and 3D models. For example, in Reference 25 the 3D geometrical model of a car, a van and a lorry is used to track vehicles in a highway. The model-based approach uses explicit *a priori* geometrical knowledge of the objects to follow, which in surveillance applications are usually people, vehicles or both. In Reference 23 the author uses two 2D models to track cars: a rectangular model for a passing car that is close to the camera and a U-shape model for the rear of the car in the distance or just in front of the camera. The system consists of an image acquisition module, a lane and car detection, a process co-ordinator and a multiple car tracker. In some multi-camera systems like [74], the focus is on extracting trajectories, which are used to build a geometric and probabilistic model for long-term prediction, and not the object itself. The *a priori* knowledge can be obtained by computing the object's appearance as a function of its position relative to the camera. The scene geometry is obtained in the same way. In order to build the shape models, the use of camera calibration techniques becomes important. A survey of different techniques for camera calibration can be found in Reference 81. Once *a priori* knowledge is available, it may be utilised in a robust tracking algorithm dealing with varying conditions such as changing illumination, offering a better performance in solving (self) occlusions or (self) collisions. It is relatively simple to create constraints in the objects' appearance model by using model-based approaches; for example, the constraint that people appear upright and in contact with the ground is commonly used in indoor and outdoor applications.

The object recognition task then becomes the process of utilising model-based techniques in an attempt to exploit such knowledge. A number of approaches can be applied to classify the new detected objects. The integrated system presented in References 53 and 25 can recognise and track vehicles using a defined 3D model of a vehicle, giving its position in the ground plane and its orientation. It can also recognise and track pedestrians using a prior 2D model silhouette shape, based on B-spline contours. A common tracking method is to use a filtering mechanism to predict each movement of the recognised object. The filter most commonly used in surveillance systems is the Kalman filter [53,73]. Fitting bounding boxes or ellipses, which are commonly called 'blobs', to image regions of maximum probability performs another tracking approach based on statistical models. In Reference 77 the author models and tracks different parts of a human body using blobs, which are described in statistical terms by a spatial and colour Gaussian distribution. In some situations of interest the assumptions made to apply linear or Gaussian filters do not hold, and then non-linear Bayesian filters, such as extended Kalman filters (EKFs) or particle filters, have been proposed. Work described in Reference 82 illustrates that in highly non-linear environments particle filters give better performance than EKFs. A particle filter is a numerical method, which weights ('particle') a representation of posterior probability densities by resampling a set of random samples associated with a weight and computing the estimate probabilities based on these weights. Then, the critical design decision using particle filters relies on the choice of importance (the initial weight) of the density function.

Another tracking approach consists in using connected components [34] to segment the changes in the scene into different objects without any prior knowledge. The approach has a good performance when the object is small, with a low-resolution approximation, and the camera placement is chosen carefully. Hidden Markov models (HMMs) have also been used for tracking purposes as presented in Reference 40, where the authors use an extension of HMMs to predict and track object trajectories. Although HMM filters are suitable for dynamic environments (because there is no assumption in the model or in the characterisation of the type of the noise as required when using Kalman filters), off-line training data are required.

Finally, due to the variability of filtering techniques used to obtain different tracking algorithms, research has been carried out on the creation of semi-automatic tools that can help create a large set of ground truth data, necessary for evaluating the performance of the tracking algorithms [63].

1.3.3   Behavioural analysis

The next stage of a surveillance system recognises and understands activities and behaviours of the tracked objects. This stage broadly corresponds to a classification problem of the time-varying feature data that are provided by the preceding stages. Therefore, it consists in matching a measured sequence to a pre-compiled library of labelled sequences that represent prototypical actions that need to be learnt by the system via training sequences. There are several approaches for matching time-varying data. Dynamic time warping (DTW) is a time-varying technique widely used in speech recognition, image pattern as in Reference 83 and recently in human movement patterns [84]. It consists of matching a test pattern with a reference pattern. Although it is a robust technique, it is now less favoured than dynamic probabilistic network models like HMMs and Bayesian networks [85,86]. The last time-varying technique that is not as widespread as HMMs, because it is less investigated for activity recognition, is neural networks (NNs). In Reference 57 the recognition of behaviours and activities is done using a declarative model to represent scenarios, and a logic-based approach to recognise pre-defined scenario models.

1.3.4 Database

One of the final stages in a surveillance system is storage and retrieval (the important aspects of user interfaces and alarm management are not considered here for lack of space). Relatively little research has been done in how to store and retrieve all the obtained surveillance information in an efficient manner, especially when it is possible to have different data formats and types of information to retrieve. In Reference 62 the authors investigate the definition and creation of data models to support the storage of different levels of abstraction of tracking data into a surveillance database. The database module is part of a multi-camera system that is presented in Figure 1.4.

In Reference 61 the authors develop a data model and a rule-based query language for video content based indexing and retrieval. Their data model allows facts

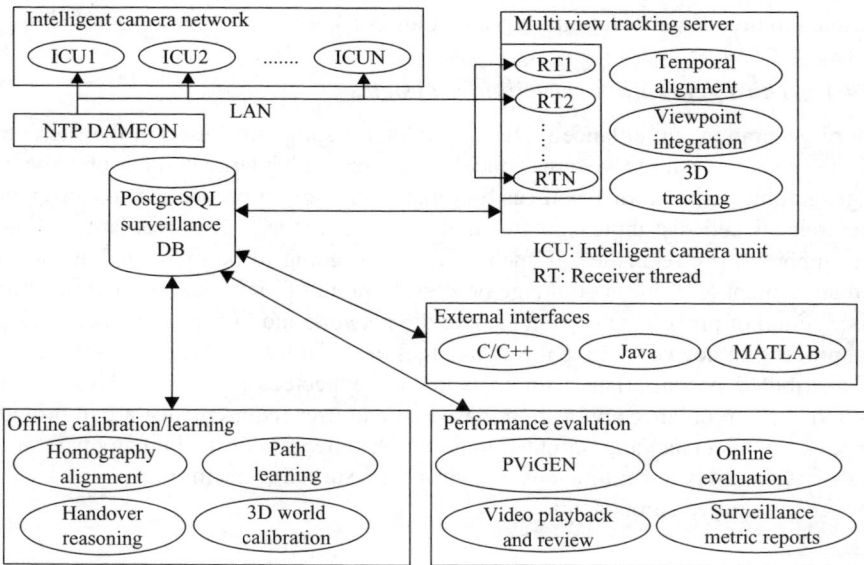

Figure 1.4   The architecture of a multi-camera surveillance system (from Makris et al. [62])

as well as objects and constraint. Retrieval is based on a rule-based query language that has declarative and operational semantics, which can be used to gather relations between information represented in the model. A video sequence is split into a set of fragments, and each fragment can be analysed to extract the information (symbolic descriptions) of interest to store into the database. In Reference 60, retrieval is performed on the basis of object classification. A stored video sequence consists of 24 frames; the last frame is the key frame that contains the information about the whole sequence. Retrieval is performed using a feature vector where each component contains information obtained from the event detection module.

1.4   Review of surveillance systems

The previous section reviewed some core computer vision techniques that are necessary for the detection and understanding of activity in the context of surveillance. It is important to highlight that the availability of a given technique or set of techniques is necessary but not sufficient to deploy a potentially large surveillance system, which implies networks of cameras and distribution of processing capacities to deal with the signals from these cameras. Therefore in this section we review what has been done to propose surveillance systems that address these requirements. The majority of the surveillance systems reviewed in this chapter are based on transport or parking lot applications. This is because most reported distributed systems tend to originate from academic research, which has tended to focus on these domains (e.g. by using university campuses for experimentation or the increasing research funding to investigate solutions in public transport).

1.4.1   Third generation surveillance systems

Third generation surveillance systems is the term sometimes used in the literature of the subject to refer to systems conceived to deal with a large number of cameras, a geographical spread of resources and many monitoring points and to mirror the hierarchical and distributed nature of the human process of surveillance. These are important pre-requisites, if such systems are going to be integrated as part of a management tool. From an image processing point of view, they are based on the distribution of processing capacities over the network and the use of embedded signal processing devices to give the advantages of scalability and robustness potential of distributed systems. The main goals that are expected of a generic third generation vision surveillance application, based on end-user requirements, are to provide good scene understanding, oriented to attract the attention of the human operator in real-time, possibly in a multi-sensor environment surveillance information and using low-cost, standard components.

1.4.2   General requirements of third generation of surveillance systems

Spatially-distributed multi-sensor environments present interesting opportunities and challenges for surveillance. Recently, there has been some investigation of data fusion techniques to cope with the sharing of information obtained from different types of sensor [41]. The communication aspects within different parts of the system play an important role with particular challenges either due to bandwidth constraints or the asymmetric nature of the communication [87].

Another relevant aspect is the security of communications between modules. For some vision surveillance systems, data might need to be sent over open networks and there are critical issues in maintaining privacy and authentication [87]. Trends in the requirements of these systems include the desirability of adding an automatic learning capability to provide the capability of characterising models of scenes to be recognised as potentially dangerous events [57,85,86,88]. A state-of-the-art survey on approaches to learning, recognising and understanding scenarios may be found in Reference 89.

1.4.3 Examples of surveillance systems

The distinction between surveillance for indoor and outdoor applications occurs because of the differences in the design at the architectural and algorithmic implementation levels. The topology of indoor environments is also different from that of outdoor environments.

Typical examples of commercial surveillance systems are DETEC [26], and Gotcha [28,29]. They are usually based on what is commonly called motion detectors, with the option of digital storage of the events detected (input images and time-stamped metadata). These events are usually triggered by objects appearing in the scene. DETEC is based on specialised hardware that allows one to connect up to 12 cameras to a single workstation. The workstation can be connected to a network and all the surveillance data can be stored in a central database available to all workstations on the network. The visualisation of the input images from the camera across Internet links is described in Reference 29.

Another example of a commercial system intended for outdoor applications is DETER [27,78] (Detection of Events for Threat Evaluation and Recognition). The architecture of the DETER system is illustrated in Figure 1.5. It is aimed at reporting unusual moving patterns of pedestrians and vehicles in outdoor environments such as car parks. The system consists of two parts: the computer vision module and the threat assessment or alarms management module. The computer vision part deals with the detection, recognition and tracking of objects across cameras. In order to do this, the system fuses the view of multiple cameras into one view and then performs the tracking of the objects. The threat assessment part consists of feature assembly or high-level semantic recognition, the off-line training and the on-line threat classifier. The system has been evaluated in a real environment by end-users, and it had a good performance in object detection and recognition. However, as pointed out in Reference 78, DETER employs a relatively small number of cameras because it is a cost-sensitive application. It is not clear whether the system has the functionality for retrieval, and even though the threat assessment has good performance, there is a lack of a feedback loop in this part that could help improve performance.

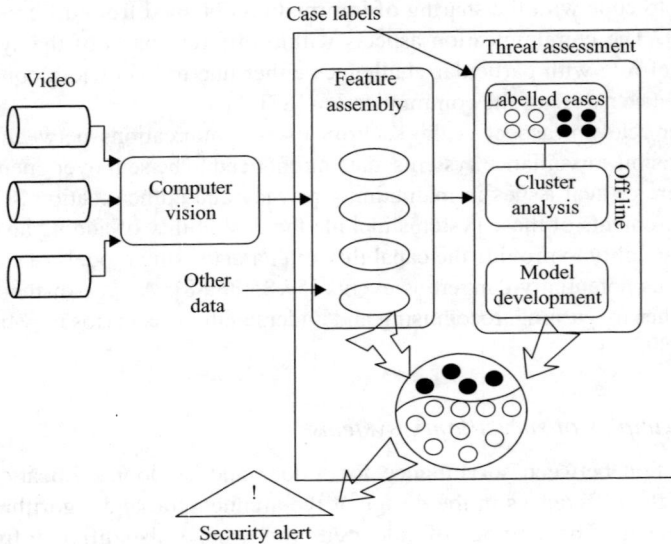

Figure 1.5   Architecture of DETER system (from Pavlidis et al. [78])

Another integrated visual surveillance for vehicles and pedestrians in parking lots is presented in Reference 53. This system presents a novel approach for dealing with interactions between objects (vehicles and pedestrians) in a hybrid tracking system. The system consists of two visual modules capable of identifying and tracking vehicles and pedestrians in a complex dynamic scene. However, this is an example of a system that considers tracking as the only surveillance task, even though the authors pointed out in Reference 53 the need for a semantic interpretation of the tracking results for scene recognition. Furthermore, a 'handover' tracking algorithm across cameras has not been established.

It is important to have a semantic interpretation of the behaviours of the recognised objects in order to build an automated surveillance system that is able to recognise and learn from the events and interactions that occur in a monitored environment. For example, in Reference 90, the authors illustrated a video-based surveillance system to monitor activities in a parking lot that performs a semantic interpretation of recognised events and interactions. The system consists of three parts: the tracker which tracks the objects and collects their movements into partial tracks; the event generator, which generates discrete events from the partial tracks according to a simple environment model; and finally, a parser that analyses the events according to a stochastic context-free grammar (SCFG) model which structurally describes possible activities. This system, like the one in Reference 53, is aimed at proving the algorithms more than at creating a surveillance system for monitoring a wide area (the system uses a single stationary camera). Furthermore, it is not clear how the system distinguishes between cars and pedestrians because the authors do not use any shape model.

In Reference 24 visual traffic surveillance for automatic identification and description of the behaviours of vehicles within parking lot scenes is presented. The system consists of a motion module, model visualisation and pose refinement, tracking and trajectory-based semantic interpretation of vehicle behaviour. The system uses a combination of colour cues and brightness information to construct the background model and applies connectivity information for pixel classification. Using camera calibration information they project the 3D model of a car onto the image plane and they use the 3D shape model-based method for pose evaluation. The tracking module is performed using EKFs. The semantic interpretation module is realised by three steps: trajectory classification, then an on-line classification step using Bayesian classifiers and finally application of natural language descriptions to the trajectory patterns of the cars that have been recognised. Although this system introduces a semantic interpretation for car behaviours, it is not clear how this system handles the interactions of several objects in the same scene at the time, and consequently the occlusions between objects. Another possible limitation is the lack of different models to represent different types of vehicle (cf. Reference 53 that includes separate 3D models for a car, van and lorry).

Other surveillance systems used in different applications (e.g. road traffic, ports and railways) can be found in [13,16,20–22]. These automatic or semi-automatic surveillance systems use more or less intelligent and robust algorithms to assist the end-user. The importance to this review of some of these systems is the illustration of how the requirements of wide geographical distribution impinge on system architecture aspects.

The author in Reference 13 expresses the need to integrate video-based surveillance systems with existing traffic control systems to develop the next generation of advanced traffic control and management systems. Most of the technologies in traffic control are based on CCTV technology linked to a control unit and in most cases for reactive manual traffic monitoring. However, there is an increasing number of CCTV systems using image processing techniques in urban road networks and highways. Therefore, the author in Reference 13 proposes to combine these systems with other existing surveillance traffic systems like surveillance systems based on networks of smart cameras. The term 'smart camera' (or 'intelligent camera') is normally used to refer to a camera that has processing capabilities (either in the same casing or nearby), so that event detection and storage of event video can be done autonomously by the camera. Thus, normally, it is only necessary to communicate with a central point when significant events occur.

Usually, integrated surveillance systems consist of a control unit system, which manages the outputs from the different surveillance systems, a surveillance signal processing unit and a central processing unit which encapsulates a vehicle ownership database. The suggestion in Reference 13 of having a control unit, which is separate from the rest of the modules, is an important aspect in the design of a third generation surveillance system. However, to survey a wide area implies geographical distribution of equipment and a hierarchical structure of the personnel who deal with security. Therefore for better scalability, usability and robustness of the system, it is desirable to have more than one control unit. Their design is likely to follow a hierarchical structure (from low-level to high-level control) that mirrors what is done in image processing where there is a differentiation between low-level and high-level processing tasks.

Continuing with traffic monitoring applications, in Reference 21 a wide-area traffic monitoring system for highway roads in Italy is presented. The system consists of two main control rooms which are situated in two different geographical places and nine peripheral control rooms which are in direct charge of road operation: toll collection, maintenance and traffic control. Most of the sensors used to control traffic are CCTV. Images are centrally collected and displayed in each peripheral control room. They have installed pan, tilt and zoom (PTZ) colour cameras in places where the efficiency of CCTV is limited, for example, by weather conditions. The system is able to detect automatically particular conditions and therefore to attract human attention. Each peripheral control room receives and manages in a multi-session environment the MPEG-1 compressed video for full motion traffic images at transmission rates of up to 2 Mbps, from each peripheral site. Therefore, there is an integration of image acquisition, coding and transmission sub-systems in each peripheral site. In some peripheral sites that control tunnels, they have a commercial sub-system that detects stopped vehicles or queues. Even though this highway traffic monitoring system is not fully automatic, it shows the importance of having a hierarchical structure of control and image processing units. Moreover, it shows the importance of coding and transmission bandwidth requirements for wide-area surveillance systems.

The authors in Reference 22 present a video-based surveillance system for measuring traffic parameters. The aim of the system is to capture video from cameras that are placed on poles or other structures looking down at traffic. Once the video is captured, digitised and processed by on-site embedded hardware, it is transmitted in summary form to a transportation management centre (TMC) for computing multi-site statistics like travel times. Instead of using 3D models of vehicles as in Reference 24 or 25, the authors use feature-based models like corners, which are tracked from entry to exit zones defined off-line by the user. Once these corner features have been tracked, they are grouped into single candidate vehicles by the sub-features grouping module. This grouping module constructs a graph over time where vertices are sub-feature tracks, edges are grouping relationships between tracks and connected components correspond to the candidate vehicle. When the last track of a connected component enters the exit region, a new candidate vehicle is generated and the component is removed from the grouping graph. The system consists of a host PC connected to a network of 13 digital signal processors (DSPs). Six of these DSPs perform the tracking, four perform the corner detection and one acts as the tracker controller. The tracker controller is connected to a DSP that is an image frame-grabber and to another DSP which acts as a display. The tracker update is sent to the host PC which runs the grouper due to memory limitations. The system has good performance not only in congested traffic conditions but also at night-time and in urban intersections.

Following the aim of Reference 22, the authors in Reference 37 develop a vision-based surveillance system for monitoring traffic flow on a road, but focusing on the detection of cyclists and pedestrians. The system consists of two main distributed processing modules: the tracking module which processes in real-time and is placed by the roadside on a pole and the analysis module which performs off-line in a PC. The tracking module consists of four tasks: motion detection, filtering, feature extraction using quasi-topological features (QTC) and tracking using first order Kalman filters. The shape and the trajectory of the recognised objects are extracted and stored in a removable memory card, which is transferred to the PC to achieve the analysis process using learning vector quantisation for producing the final counting. This system has some shortcomings. The image algorithms are not robust enough (the background model is not robust enough to cope with changing conditions or shadows) and depend on the position of the camera. The second problem is that even though tracking is performed in real time, the analysis is performed off-line, and therefore it is not possible to do flow statistics or monitoring in real-time.

In Reference 16 the architecture of a system for surveillance in a maritime port is presented. The system consists of two sub-systems: image acquisition and visualisation. The architecture is based on a client/server design. The image acquisition sub-system has video server module, which can handle four cameras at the same time. This module acquires the images from camera streams, which are compressed, and then the module broadcasts the compressed images to the network using TCP/IP and at the same time records the images on hard disks. The visualisation module is performed by client sub-systems, which are based on PC boards. This module allows the selection of any camera using a pre-configured map and the configuration of the video server. Using an Internet server module it is possible to display the images through the Internet. The system is claimed to have the capability of supporting more than 100 cameras and 100 client stations at the same time, even though the reported implementation had 24 cameras installed mainly at the gates of the port. This is an example of a simple video surveillance system (with no image interpretation), which only consists of image acquisition, distribution and display. The interesting point in this system is to see the use of a client and server architecture to deal with the distribution of the multiple digital images. Moreover, the acquisition and visualisation modules have been encapsulated in a way such that scalability of the system can be accomplished in a straightforward way, by integrating modules into the system in a 'drop' operation.

In Reference 20 a railway station CCTV surveillance system in Italy is presented. As in Reference 21, the system has a hierarchical structure distributed between main (central) control rooms and peripheral site (station) control rooms. The tasks that are performed in the central control room are acquisition and display of the live or recorded images. The system also allows the acquisition of images from all the station control rooms through communication links and through specific coding and decoding devices. Digital recording, storage and retrieval of the image sequences as well as the selection of specific CCTV camera and the deactivation of the alarm system are carried out in the central room. The main tasks performed in each station control room are acquisition of the images from the local station CCTV cameras and linking with the central control room to transmit the acquired or archived images in real time and to receive configuration procedures. The station control room also handles the transmission of an image of a specific CCTV camera at higher rate under request or automatically when an alarm has been raised. The management and deactivation of local alarms is handled from the station control room. Apart from the central control room and the station control rooms, there is a crisis room for the management of railway emergencies. Although this system presents a semi-automatic, hierarchical and distributed surveillance system, the role played by human operators is still central because there is no processing (object recognition or motion estimation) to channel the attention of the monitoring personnel.

Ideally, a third generation of surveillance system for public transport applications would provide a high level of automation in the management of information as well as alarms and emergencies. That is the stated aim of the following two surveillance systems research projects (other projects in public transportation that are not included here can be found in Reference 47).

CROMATICA [19] (crowd management with telematic imaging and communication assistance) was an EU-funded project whose main goal was to improve the surveillance of passengers in public transport, enabling the use and integration of technologies like video-based detection and wireless transmission. This was followed by another EU-funded project called PRISMATICA [12] (pro-active integrated systems for security management by technological institutional and communication assistance) that looked at social, ethical, organisational and technical aspects of surveillance for public transport. A main technical output was a distributed surveillance system. It is not only a wide-area video-based distributed system like ADVISOR (annotated digital video for intelligent surveillance and optimised retrieval) [18], but it is also a wide-area multi-sensor distributed system, receiving inputs from CCTV, local wireless camera networks, smart cards and audio sensors. PRISMATICA then consists of a network of intelligent devices (that process sensor inputs) that send and receive messages to/from a central server module (called 'MIPSA') that co-ordinates device activity, archives/retrieves data and provides the interface with a human operator. Figure 1.6 shows the architecture of PRISMATICA. Like ADVISOR (see below), PRISMATICA is a modular and scalable architecture approach using standard commercial hardware.

ADVISOR was also developed as part of an EU-funded project. It aims to assist human operators by automatic selection, recording and annotation of images that have events of interest. In other words, ADVISOR interprets shapes and movements in scenes being viewed by the CCTV to build up a picture of the behaviour of people in the scene. ADVISOR stores all video output from cameras. In parallel with recording video information, the archive function stores commentaries (annotations) of events detected in particular sequences. The archive video can be searched using queries for the annotation data or according to specific times. Retrieval of video sequences can take place alongside continuous recording. ADVISOR is intended to be an open and scalable architecture approach and is implemented using standard commercial hardware with an interface to a wide-bandwidth video distribution network. Figure 1.7 shows a possible architecture of the ADVISOR system. It consists of a network of ADVISOR units, each of which is installed in a different underground station and consists of an object detection and recognition module, tracking module and behavioural analysis and database module.

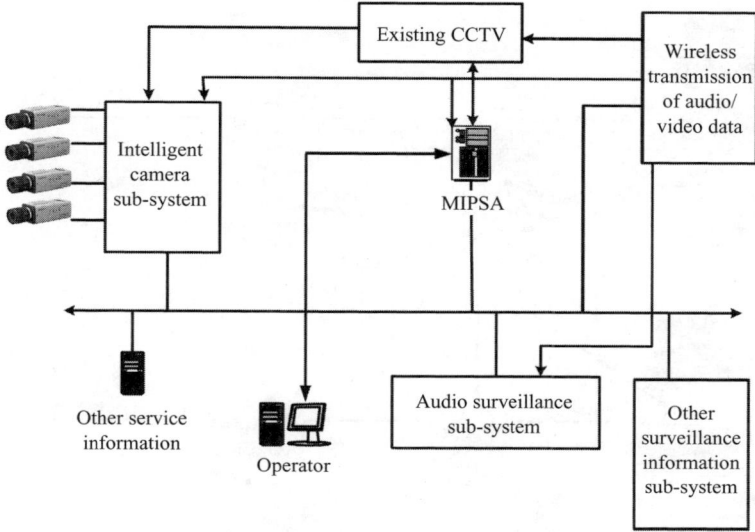

Figure 1.6 Architecture of PRISMATICA system (from Ping Lai Lo et al. [12])

Although both systems are classified as distributed architectures, they have a significant main difference in that PRISMATICA employs a centralised approach whereas ADVISOR can be considered as a semi-distributed architecture. PRISMATICA is built with the concept of a main or central computer which controls and supervises the whole system. This server thus becomes a critical single point of failure for the whole system. ADVISOR can be seen as a network of independent dedicated processor nodes (ADVISOR units), avoiding single point-of-failure at first sight. Nevertheless, each node is a rack with more than one CPU and each node contains a central computer, which controls the whole node, that still depends on a main central server which sets up the controls in each node. Moreover, the number of CPUs in each node is directly proportional to the number of existing image processing modules, making the system difficult to scale and hard to build in cost-sensitive applications.

In Reference 91 the authors report the design of a surveillance system with no server to avoid this centralisation, making all the independent sub-systems completely self-contained and then setting up all these nodes to communicate with each other without having a mutually shared communication point. This approach avoids the disadvantages of the centralised server and moves all the processes directly to the camera, making the system a group of smart cameras connected across the network. The fusion of information between 'crunchers' (as they are referred to in the article) is done through a defined protocol, after the configuration of the network of smart cameras or 'crunchers'. The defined protocol has been validated with a specific verification tool called spin. The format of the information to share between 'crunchers' is based on a common data structure or object model with different stages depending,

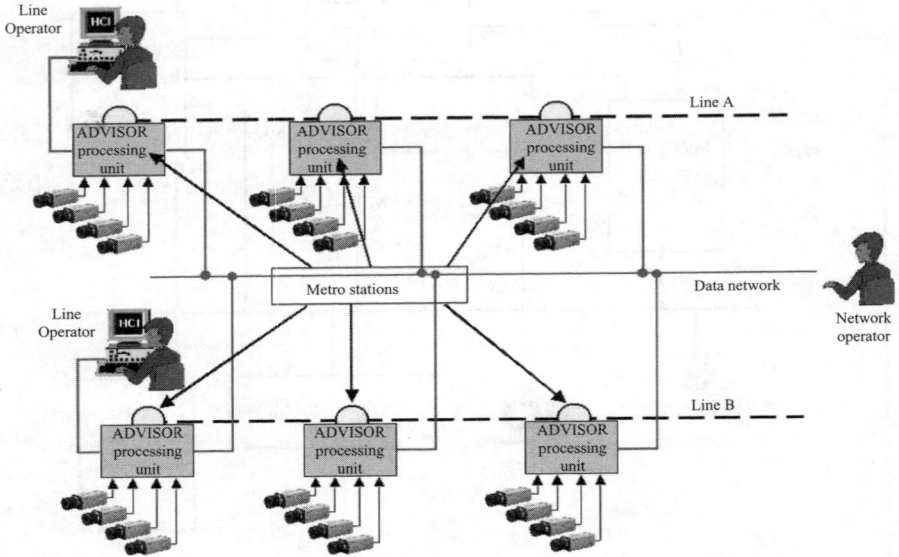

Figure 1.7   Proposed architecture of ADVISOR system (from Reference 18). Dashed black lines correspond to metro railway and solid lines correspond to computer links

for example, if the object is recognised or is migrating from the field of view of one camera to another. However, the approach to distributed design is to build using specific commercial embedded hardware (EVS units). These embedded units consist of a camera, processor, frame grabber, network adapter and database. Therefore, in cost-sensitive applications where a large number of cameras are required, this approach might be unsuitable.

As part of the VSAM project, Reference 76 presents a multi-camera surveillance system following the same idea as in Reference 92, that is, the creation of a network of 'smart' sensors that are independent and autonomous vision modules. Nevertheless, in Reference 76, these sensors are capable of detecting and tracking objects, classifying the moving objects into semantic categories such as 'human' or 'vehicle' and identifying simple human movements such as walking. In Reference 92 the smart sensors are only able to detect and track moving objects. Moreover, the algorithms in Reference 92 are based on indoor applications. Furthermore, in Reference 76 the user can interact with the system. To achieve this interactivity, there are system-level algorithms which fuse sensor data, perform the processing tasks and display the results in a comprehensible manner. The system consists of a central control unit (OCU) which receives the information from multiple independent remote processing units (SPUs). The OCU interfaces with the user through a GUI module.

Monitoring wide areas requires the use of a significant number of cameras to cover as much area as possible and to achieve good performance in the automatic surveillance operation. Therefore, the need to co-ordinate information across cameras becomes an important issue. Current research points towards developing surveillance systems that consist of a network of cameras (monocular, stereo, static or PTZ) which perform the type of vision algorithms that we have reviewed earlier, but also using information from neighbouring cameras. The following sections highlight the main work in this field.

1.4.3.1 Co-operative camera systems

An application of surveillance of human activities for sports application is presented in Reference 35. The system consists of eight cameras, eight feature server processes and a multi-tracker viewer. Only the cameras are installed on the playing area, and the raw images are sent through optical fibres to each feature server module. Each module realises the segmentation, the single-view tracking and the object classification and sends the results to the multi-tracker module, which merges all the information from the single-view trackers using a nearest neighbour method based on the Mahalanobis distance.

CCN (co-operative camera network) [27] is an indoor application surveillance system that consists of a network of nodes. Each node is composed of a PTZ camera connected to a PC and a central console to be used by the human operator. The system reports the presence of a visually tagged individual inside the building by assuming that human traffic is sparse (an assumption that becomes less valid as crowd levels increase). Its purpose is to monitor potential shoplifters in department stores.

In Reference 33 a surveillance system for a parking lot application is described. The architecture of the system consists of one or more static camera sub-systems (SCSs) and one or more active camera sub-systems (ACSs). First, the target is detected and tracked by the static sub-systems; once the target has been selected a PTZ, which forms the ACS, is activated to capture high-resolution video of the target. The data fusion for the multi-tracker is done using the Mahalanobis distance. Kalman filters are used for tracking, as in Reference 35.

In Reference 36 the authors present a multi-camera tracking system that is included in an intelligent environment system called 'EasyLiving' which aims at assisting the occupants of that environment by understanding their behaviour. The multi-camera tracking system consists of two sets of stereo cameras (each set has three small colour cameras). Each set is connected to a PC that runs the 'stereo module'. The two stereo modules are connected to a PC which runs the tracker module. The output of the tracker module is the localisation and identity of the people in the room. This identity does not correspond to the natural identity of the person but to an internal temporary identity which is generated for each person using a colour histogram that is provided by the stereo module each time. The authors use the depth and the colour information provided from the cameras to apply background subtraction and to allocate 3D blobs, which are merged into person shapes by clustering regions. Each stereo module reports the 2D ground plane locations of its person blobs to the tracking module. Then, the tracker module uses knowledge of the relative locations of the cameras, field of view and heuristics of the movement of people to produce the locations and identities of the people in the room. The performance of the tracking system is good when there are fewer than three people in the room and when the people wear different colour outfits; otherwise, due to the poor clustering results, performance is reduced drastically.

In Reference 92 an intelligent video-based visual surveillance system (IVSS) is presented which aims to enhance security by detecting certain types of intrusion in dynamic scenes. The system involves object detection and recognition (pedestrians and vehicles) and tracking. The system is based on a distribution of a static multi-camera monitoring module via a local area network. The design architecture of the system is similar to that of ADVISOR [18], and the system consists of one or more clients plus a server, which are connected through TCP/IP. The clients only connect to the server (and not to other clients), while the server talks with all clients. Therefore there is no data fusion across cameras. The vision algorithms are developed in two stages: hypothesis generation (HG) and hypothesis verification (HV). The first stage realises a simple background subtraction. The second stage compensates the non-robust background subtraction model. This stage is essentially a pattern classification problem and it uses a Gabor filter to extract features, for example, the strong edges and lines at different orientation of the vehicles and pedestrians, and SVMs to perform the classifications. Although this is an approach to developing a distributed surveillance system, there is no attempt at fusing information across cameras. Therefore, it is not possible to track objects across clients. Furthermore, the vision algorithms do not include activity recognition, and although the authors claim to compensate the simple motion detection algorithm using the Gabor filters, it is not clear how these filters and SVMs cope with uncertainties in the tracking stage, for example, occlusions or shadows.

In Reference 72 a multi-camera surveillance system for face detection is illus-trated. The system consists of two cameras (one of the cameras is a CCD pan-tilt-zoom and the other one is a remote control camera). The system architecture is based on three main modules using a client/server approach as a solution for the distribution. The three modules are sensor control, data fusion and image processing. The sensor control module is a dedicated unit to control directly the two cameras and the infor-mation that flows between them. The data fusion module controls the position of the remote control camera depending on the inputs received from the image processing and sensor control module. It is interesting to see how the authors use the information obtained from the static camera (the position of the recognised object) to feed the other camera. Therefore, the remote control camera can zoom to the recognised human to detect the face.

An interesting example of a multi-tracking camera surveillance system for indoor environments is presented in Reference 73. The system is a network of camera pro-cessing modules, each of which consists of a camera connected to a computer, and a control module, which is a PC that maintains the database of the current objects in the scene. Each camera processing module realises the tracking process using Kalman filters. The authors develop an algorithm which divides the tracking task between the cameras by assigning the tracking to the camera which has better visibility of the object, taking into account the occlusions. This algorithm is implemented in the control module. In this way, unnecessary processing is reduced. Also, it makes it possible to solve some occlusion problems in the tracker by switching from one camera to another camera when the object is not visible enough. The idea is interesting because it shows a technique that exploits distributed processing to improve detection performance. Nevertheless, the way that the algorithm decides which camera is more appropriate is performed using a 'quality service of tracking' function. This function is defined based on the sizes of the objects in the image, estimated from the Kalman filter, and the object occlusion status. Consequently, in order to calculate the size of the object with respect to the camera, all the cameras have to try to track the object. Moreover, the system has been built with the constraint that all the cameras have overlapping views (if there were topographic knowledge of the cameras the calculation of this function could be applied only to the cameras which have overlapping views). Furthermore, in zones where there is a gap between views, the quality service of tracking function would drop to zero, and if the object reappears it would be tracked as a new object.

VIGILANT [32] is a multi-camera surveillance system (Figure 1.8) which monitors pedestrians walking in a parking lot. The system tracks people across cameras using software agents. For each detected person in each camera an agent is created to hold the information. The agents communicate to obtain a consensus decision of whether or not they are assigned the same person who is being seen from different cameras by reasoning on trajectory geometry in the ground plane.

As has been illustrated, in a distributed multi-camera surveillance system it is important to know the topology of the links between the cameras that make up the system in order to recognise, understand and follow an event that may be captured on one camera and to follow it in other cameras. Most of the multi-camera systems that have been discussed in this review use a calibration method to compute the network camera topology. Moreover, most of these systems try to combine the tracks of the same target that are simultaneously visible in different camera views.

In Reference 62 the authors present a distributed multi-camera tracking surveillance system for outdoor environments (its architecture can be seen in Figure 1.4).

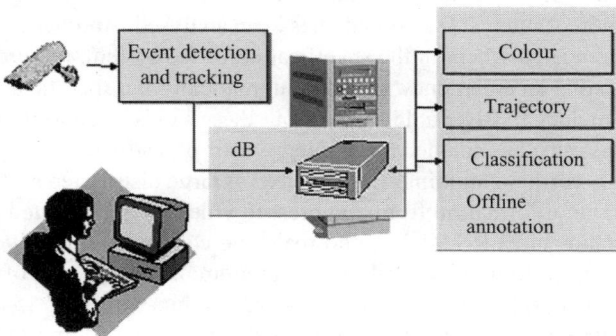

Figure 1.8   Architecture of VIGILANT system (from Greenhill et al. [32])

An approach is presented which is based on learning a probabilistic model of an activity in order to establish links between camera views in a correspondence-free manner. The approach can be used to calibrate the network of cameras and does not require correspondence information. The method correlates the number of incoming and outgoing targets for each camera view, through detected entry and exit points. The entry and exit zones are modelled by a GMM and initially these zones are learnt automatically from a database using an EM algorithm. This approach provides two main advantages: no previous calibration method is required and the system allows tracking of targets across the 'blind' regions between camera views. The first advantage is particularly useful because of the otherwise resource-consuming process of camera calibration for wide-area distributed multi-camera surveillance systems with a large number of cameras [18,20,21,47].

1.5   Distribution, communication and system design

In Section 1.3 we considered different techniques that have been used to develop more robust and adaptive algorithms. In Section 1.4 we presented a review of different architectures of distributed surveillance systems. Although the design of some of these systems can look impressive, there are some aspects where it will be advantageous to dedicate more attention for the development of distributed surveillance systems for the next few years. These include the distribution of processing tasks, the use of new technologies as well as the creation of metadata standards or new protocols to cope with current limitations in bandwidth capacities. Other aspects that should be taken into consideration for the next generation of surveillance systems are the design of scheduling control and more robust and adaptive algorithms. A field that needs further research is that of alarm management, which is an important part of an automatic surveillance system, for example, when different priorities and goals need to be considered. For example, in Reference 93 the authors describe work carried out in a robotics field, where the robot is able to focus attention in a certain region of interest, extract its features and recognise objects in the region. The control part of the system allows the robot to refocus its attention in a different region of interest and skip a region of interest that already has been analysed. Another example can be found in Reference 18 where in the specification of the system, requirements of the system like 'to dial an emergency number automatically if a specific alarm has been detected' are included. To be able to carry out these kinds of actions command and control systems must be included as an integral part of a surveillance system.

Other work worth mentioning in the context of large distributed systems has considered extracting information from compressed video [65], dedicated protocols for distributed architectures [69,94,95], and real-time communications [96]. Work has also been conducted to build an embedded autonomous unit as part of a distributed architecture [30,68,91]. Several researchers are dealing with PTZ [54,72] because this kind of camera can survey wider areas and can interact in more efficient ways with the end-user who can zoom when necessary. It is also important to incorporate scheduling policies to control resource allocation as illustrated in Reference 97.

Work in multiple robot systems [98] illustrates how limited communications bandwidth affects robot performance and how this performance is linked to the number of robots that share the bandwidth. A similar idea is presented in References 71 and 99 for surveillance systems, while in Reference 94 an overview of the state-of-the-art of multimedia communication technologies and a standard are presented. On the whole, the work on intelligent distributed surveillance systems has been led by computer vision laboratories perhaps at the expense of system engineering issues. It is essential to create a framework or methodology for designing distributed wide-area surveillance systems, from the generation of requirements to the creation of design models by defining functional and intercommunication models as is done in the creation of distributed concurrent real-time systems in other disciplines like control systems in aerospace. Therefore, as has been mentioned earlier in this chapter, in the future the realisation of a wide-area distributed intelligent surveillance system should be through a combination of different disciplines, computer vision, telecommunications and system engineering being clearly needed. Work related to the development of a design framework for developing video surveillance systems can be found in [91,99,100]. Distributed virtual applications are discussed in Reference 101 and embedded architectures in Reference 102. For example, much could be borrowed from the field of autonomous robotic systems on the use of multi-agents, where non-centralised collections of relatively autonomous entities interact with each other in a dynamic environment. In a surveillance system, one of the principal costs is the sensor suite and payload. A distributed multi-agent approach may offer several advantages. First, intelligent co-operation between agents may allow the use of less expensive sensors and therefore a larger number of sensors may be deployed over a greater area. Second, robustness is increased, since even if some agents fail, others remain to perform the mission. Third, performance is more flexible, and there is a distribution of tasks at various locations between groups of agents. For example, the likelihood of correctly classifying an object or target increases if multiple sensors are focused on it from different locations.

1.6 Conclusions

This chapter has presented the state of development of intelligent distributed surveillance systems, including a review of current image processing techniques that are used in different modules that constitute part of surveillance systems. Looking at these image processing tasks, it has identified research areas that need to be investigated further such as adaptation, data fusion and tracking methods in a co-operative multi-sensor environment and extension of techniques to classify complex activities and interactions between detected objects. In terms of communication or integration between different modules it is necessary to study new communication protocols and the creation of metadata standards. It is also important to consider improved means of task distribution that optimise the use of central, remote facilities and data communication networks. Moreover, one of the aspects that the authors believe is essential in the coming future for the development of distributed surveillance systems is the definition of a framework to design distributed architectures firmly rooted on systems engineering best practice, as used in other disciplines such as control aerospace systems.

The growing demand for safety and security has led to more research in building more efficient and intelligent automated surveillance systems. Therefore, a future challenge is to develop a wide-area distributed multi-sensor surveillance system which has robust, real-time computer algorithms able to perform with minimal manual reconfiguration on variable applications. Such systems should be adaptable enough to adjust automatically and cope with changes in the environment like lighting, scene geometry or scene activity. The system should be extensible enough, be based on standard hardware and exploit plug-and-play technology.

Acknowledgements

This work was part of the EPSRC-funded project COHERENT (computational heterogeneously timed networks); the grant number is GR/R32895 (http://async.org.uk/coherent/). We would like to thank Mr David Fraser and Prof Tony Davies for their valuable observations.

References

1 First IEEE Workshop on Visual Surveillance, January 1998, Bombay, India, ISBN 0-8186-8320-1.
2 Second IEEE Workshop on Visual Surveillance, January 1999, Fort Collins, Colorado, ISBN 0-7695-0037-4.
3 Third IEEE International Workshop on Visual Surveillance (VS'2000), July 2000, Dublin, Ireland, ISBN 0-7695-0698-4.
4 First IEE Workshop on Intelligent Distributed Surveillance Systems, February 2003, London, ISSN 0963-3308.
5 Second IEE Workshop on Intelligent Distributed Surveillance Systems, February 2004, London, ISBN 0-86341-3927.
6 IEEE Conference on Advanced Video and Signal Based Surveillance, July 2003, ISBN 0-7695-1971-7.
7 Special issue on visual surveillance, *International Journal of Computer Vision*, June 2000.
8 Special issue on visual surveillance, *IEEE Transactions on Pattern Analysis and Machine Intelligence*, August 2000, ISSN 0162-8828/00.
9 Special issue on third generation surveillance systems, *Proceedings of IEEE*, October 2001, ISBN 0018-9219/01.
10 Special issue on human motion analysis, *Computer Vision and Image Understanding*, March 2001.
11 www.cieffe.com.

12 B. Ping Lai Lo, J. Sun, and S.A. Velastin. Fusing visual and audio information in a distributed intelligent surveillance system for public transport systems. *Acta Automatica Sinica*, 2003;29(3):393–407.

13 C. Nwagboso. User focused surveillance systems integration for intelligent transport systems. In Advanced Video-Based Surveillance Systems, C.S. Regazzoni, G. Fabri, and G. Vernazza, Eds. Kluwer Academic Publishers, Boston, 1998, pp. 8–12.

14 www.sensis.com/docs/128.

15 M.E. Weber and M.L. Stone. Low altitude wind shear detection using airport surveillance radars. *IEEE Aerospace and Electronic Systems Magazine*, 1995; 10(6):3–9.

16 A. Pozzobon, G. Sciutto, and V. Recagno. Security in ports: the user requirements for surveillance systems. In Advanced Video-Based Surveillance Systems, C.S. Regazzoni, G. Fabri, and G. Vernazza, Eds. Kluwer Academic Publishers, Boston, 1998, pp. 18–26.

17 P. Avis. Surveillance and Canadian maritime domestic security. *Canadian Military Journal*, 4(Spring 2003):9–15, ISSN 1492-0786.

18 ADVISOR specification documents (internal classification 2001).

19 http://dilnxsvr.king.ac.uk/cromatica/.

20 N. Ronetti and C. Dambra. Railway station surveillance: the Italian case. In Multimedia Video Based Surveillance Systems, G.L. Foresti, P. Mahonen, and C.S. Regazzoni, Eds. Kluwer Academic Publishers, Boston, 2000, pp. 13–20.

21 M. Pellegrini and P. Tonani. Highway traffic monitoring. In Advanced Video-Based Surveillance Systems, C.S. Regazzoni, G. Fabri, and G. Vernazza, Eds. Kluwer Academic Publishers, Boston, 1998, pp. 27–33.

22 D. Beymer, P. McLauchlan, B. Coifman, and J. Malik. A real-time computer vision system for measuring traffic parameters. In *Proceedings of the 1997 Conference on Computer Vision and Pattern Recognition*, IEEE Computer Society, pp. 495–502.

23 Z. Zhi-Hong. Lane detection and car tracking on the highway. *Acta Automatica Sinica*, 2003;29(3):450–456.

24 L. Jian-Guang, L. Qi-Feing, T. Tie-Niu, and H. Wei-Ming. 3-D model based visual traffic surveillance. *Acta Automatica Sinica*, 2003;29(3):434–449.

25 J.M. Ferryman, S.J. Maybank, and A.D. Worrall. Visual surveillance for moving vehicles. *International Journal of Computer Vision*, 37(2):187–197, Kluwer Academic Publishers, Netherlands, 2000.

26 http://www.detec.no.

27 I. Paulidis and V. Morellas. Two examples of indoor and outdoor surveillance systems. In Video-Based Surveillance Systems, P. Remagnino, G.A. Jones, N. Paragios, and C.S. Regazzoni, Eds. Kluwer Academic Publishers, Boston, 2002, pp. 39–51.

28 http://www.gotchanow.com.

29 secure30.softcomca.com/fge_biz.

30 T. Brodsky, R. Cohen, E. Cohen-Solal *et al.* Visual surveillance in retail stores and in the home. In Advanced Video-Based Surveillance Systems. Kluwer Academic Publishers, Boston, 2001, pp. 50–61.

31 R. Cucchiara, C. Grana, A. Patri, G. Tardini, and R. Vezzani. Using computer vision techniques for dangerous situation detection in domotics applications. In *Proceedings the 2nd IEE Workshop on Intelligent Distributed Surveillance Systems*, London, 2004, pp. 1–5, ISBN 0-86341-3927.

32 D. Greenhill, P. Remagnino, and G.A. Jones. VIGILANT: content-querying of video surveillance streams. In Video-Based Surveillance Systems, P. Remagnino, G.A. Jones, N. Paragios, and C.S. Regazzoni, Eds. Kluwer Academic Publishers, Boston, USA, 2002, pp. 193–205.

33 C. Micheloni, G.L. Foresti, and L. Snidaro. A co-operative multi-camera system for video-surveillance of parking lots. *Intelligent Distributed Surveillance Systems Symposium by the IEE*, London, 2003, pp. 21–24, ISSN 0963-3308.

34 T.E. Boult, R.J. Micheals, X. Gao, and M. Eckmann. Into the woods: visual surveillance of non-cooperative and camouflaged targets in complex outdoor settings. *Proceedings of the IEEE*, 2001;89(1):1382–1401.

35 M. Xu, L. Lowey, and J. Orwell. Architecture and algorithms for tracking football players with multiple cameras. In *Proceedings of the 2nd IEE Workshop on Intelligent Distributed Surveillance Systems*, London, 2004, pp. 51–56, ISBN: 0-86341-3927.

36 J. Krumm, S. Harris, B. Meyers, B. Brumit, M. Hale, and S. Shafer. Multi-camera multi-person tracking for easy living. In *Third IEEE International Workshop on Visual Surveillance*, Ireland, 2000, pp. 8–11, ISBN 0-7695-0698-04.

37 J. Heikkila and O. Silven. A real-time system for monitoring of cyclists and pedestrians. In *Second IEEE International Workshop on Visual Surveillance*, Colorado, 1999, pp. 74–81, ISBN 0-7695-0037-4.

38 I. Haritaoglu, D. Harwood, and L.S. Davis. W^4: real-time surveillance of people and their activities. *IEEE Transactions on Pattern Analysis and Machine Intelligence*, 2000;22(8):809–830.

39 Z. Geradts and J. Bijhold. Forensic video investigation. In Multimedia Video Based Surveillance Systems, G.L. Foresti, P. Mahonen, and C.S. Regazzoni, Eds. Kluwer Academic Publishers, Boston, 2000, pp. 3–12.

40 H. Hai Bui, S. Venkatesh, and G.A.W. West. Tracking and surveillance in wide-area spatial environments using the abstract hidden Markov model. *IJPRAI*, 2001;15(1):177–195.

41 R.T. Collins, A.J. Lipton, T. Kanade *et al. A System for Video Surveillance and Monitoring.* Robotics Institute, Carnegie Mellon University, 2000, pp. 1–68.

42 www.objectvideo.com.

43 www.nice.com.

44 www.pi-vision.com.

45 www.ipsotek.com.

46 www.neurodynamics.com.

47 S.A. Velastin. *Getting the Best Use out of CCTV in the Railways*, Rail Safety and Standards Board, July 2003, pp. 1–17. Project reference: 07-T061.

48 I. Haritaoglu, D. Harwood, and L.S. Davis. Hydra: multiple people detection and tracking using silhouettes. In *Proceedings of IEEE International Workshop Visual Surveillance*, 1999, pp. 6–14, ISBN 0-7695-0037-4.

49 J. Batista, P. Peixoto, and H. Araujo. Real-time active visual surveillance by integrating. In *Workshop on Visual Surveillance*, India, 1998, pp. 18–26.

50 Y.A. Ivanov, A.F. Bobick, and J. Liu. Fast lighting independent background. *International Journal of Computer Vision*, 2000;37(2):199–207.

51 R. Pless, T. Brodsky, and Y. Aloimonos. Detecting independent motion: the statics of temporal continuity. *IEEE Transactions on Pattern Analysis and Machine Intelligence*, 2000, pp. 768–773.

52 L.C. Liu, J.-C. Chien, H.Y.-H. Chuang, and C.C. Li. A frame-level FSBM motion estimation architecture with large search range. In *IEEE Conference on Advanced Video and Signal Based Surveillance*, Florida, 2003, pp. 327–334, ISBN 0-7695-1971-7.

53 P. Remagnino, A. Baumberg, T. Grove, D. Hogg, T. Tan, A. Worral, and K. Baker. An integrated traffic and pedestrian model-based vision system. *BMVC97 Proceedings*, Israel, vol. 2, pp. 380–389, ISBN 0-9521-8989-5.

54 K.C. Ng, H. Ishiguro, M. Trivedi, and T. Sogo. Monitoring dynamically changing environments by ubiquitous vision system. In *Second IEEE Workshop on Visual Surveillance*, Colorado, 1999, pp. 67–74, ISBN 0-7695-0037-4.

55 J. Orwell, P. Remagnino, and G.A. Jones. Multicamera color tracking. In *Second IEEE Workshop on Visual Surveillance*, Colorado, 1999, pp. 14–22, ISBN 0-7695-0037-4.

56 T. Darrell, G. Gordon, J. Woodfill, H. Baker, and M. Harville. Robust, real-time people tracking in open environments using integrated stereo, color, and face detection. In *IEEE Workshop on Visual Surveillance*, India, 1998, pp. 26–33, ISBN 0-8186-8320-1.

57 N. Rota and M. Thonnat. Video sequence interpretation for visual surveillance. In *Third IEEE International Workshop on Visual Surveillance*, Dublin, 2000, pp. 59–68, ISBN 0-7695-0698-4.

58 J. Owens and A. Hunter. Application of the self-organising map to trajectory classification. In *Third IEEE International Workshop on Visual Surveillance*, Dublin, 2000, pp.77–85, ISBN 0-7695-0698-4.

59 C. Stauffer, W. Eric, and L. Grimson. Learning patterns of activity using real-time tracking. *IEEE Transactions on Pattern Analysis and Machine Intelligence*, 2000;22(8):747–757.

60 E. Stringa and C.S. Regazzoni. Content-based retrieval and real-time detection from video sequences acquired by surveillance systems. In *International Conference on Image Processing*, Chicago, 1998, pp. 138–142.

61 C. Decleir, M.-S. Hacid, and J. Kouloourndijan. A database approach for modelling and querying video data. In *Proceedings of the 15th International Conference on Data Engineering*, Australia, 1999, pp. 1–22.

62 D. Makris, T. Ellis, and J. Black. Bridging the gaps between cameras. In *International Conference on Multimedia and Expo*, Taiwan, June 2004.

63 J. Black, T. Ellis, and P. Rosin. A novel method for video tracking performance evaluation. In *The Joint IEEE International Workshop on Visual Surveillance and Performance Evaluation of Tracking and Surveillance*, October, France, 2003, pp. 125–132, ISBN 0-7695-2022-7.

64 D.M. Gavrila. The analysis of human motion and its application for visual surveillance. *Computer Vision and Image Understanding*, 1999;73(1):82–98.

65 P. Norhashimah, H. Fang, and J. Jiang. Video extraction in compressed domain. In *IEEE Conference on Advanced Video and Signal Based Surveillance*, Florida, 2003, pp. 321–327, ISBN 0-7695-1971-7.

66 F. Soldatini, P. Mähönen, M. Saaranen, and C.S. Regazzoni. Network management within an architecture for distributed hierarchical digital surveillance systems. In Multimedia Video Based Surveillance Systems, G.L. Foresti, P. Mahonen, and C.S. Regazzoni, Eds. Kluwer Academic Publishers, Boston, 2000, pp. 143–157.

67 L.-C. Liu, J.-C. Chien, H.Y.-H. Chuang, and C.C. Li. A frame-level FSBM motion estimation architecture with large search range. In *IEEE Conference on Advanced Video and Signal Based Surveillance*, Florida, 2003, pp. 327–334.

68 A. Saad and D. Smith. An IEEE 1394-firewire-based embedded video system for surveillance applications. In *IEEE Conference on Advanced Video and Signal Based Surveillance*, Florida, 2003, pp. 213–219, ISBN 0-7695-1971-7.

69 H. Ye, Gregory C. Walsh, and L.G. Bushnell. Real-time mixed-traffic wireless networks. *IEEE Transactions on Industrial Electronics*, 2001;48(5):883–890.

70 J. Huang, C. Krasic, J. Walpole, and W. Feng. Adaptive live video streaming by priority drop. *IEEE Conference on Advanced Video and Signal Based Surveillance*, Florida, 2003, pp. 342–348, ISBN 0-7695-1971-7.

71 L. Marcenaro, F. Oberti, G.L. Foresti, and C.S. Regazzoni. Distributed architectures and logical-task decomposition in multimedia surveillance systems. *Proceedings of the IEEE*, 2001;89(10):1419–1438.

72 L. Marchesotti, A. Messina, L. Marcenaro, and C.S. Regazzoni. A cooperative multisensor system for face detection in video surveillance applications. *Acta Automatica Sinica*, 2003;29(3):423–433.

73 N.T. Nguyen, S. Venkatesh, G. West, and H.H. Bui. Multiple camera coordination in a surveillance system. *Acta Automatica Sinica*, 2003;29(3):408–421.

74 C. Jaynes. Multi-view calibration from planar motion for video surveillance. In *Second IEEE International Workshop on Visual Surveillance*, Colorado, 1999, pp. 59–67, ISBN 0-7695-0037-4.

75 L. Snidaro, R. Niu, P.K. Varshney, and G.L. Foresti. Automatic camera selection and fusion for outdoor surveillance under changing weather conditions. In *IEEE Conference on Advanced Video and Signal Based Surveillance*, Florida, 2003, pp. 364–370, ISBN 0-7695-1971-7.

76 R.T. Collins, A.J. Lipton, H. Fujiyoshi, and T. Kanade. Algorithms for cooperative multisensor surveillance. *Proceedings of the IEEE*, 2001;89(10):1456–1475.

77 C. Wren, A. Azarbayejani, T. Darrell, and A. Pentland. Pfinder: real-time tracking of the human body. *IEEE Transactions on Pattern Analysis and Machine Intelligence*, 1997;19(7):780–785.

78 I. Pavlidis, V. Morellas, P. Tsiamyrtzis, and S. Harp. Urban surveillance systems: from the laboratory to the commercial world. In *Proceedings of the IEEE*, 2001;89(10):1478–1495.

79 M. Bennewitz, W. Burgard, and S. Thrun. Using EM to learn motion behaviours of persons with mobile robots. In *Proceedings of the International Conference on Intelligent Robots and Systems (IROS)*, Switzerland, 2002, pp. 502–507.

80 M. Oren, C. Papageorgiou, P. Sinham, E. Osuna, and T. Poggio. Pedestrian detection using wavelet templates. In *Proceedings of IEEE Conference on Computer Vision and Pattern Recognition*, Puerto Rico, 1997, pp. 193–199, ISSN 1063-6919/97.

81 E.E. Hemayed. A survey of self-camera calibration. In *Proceedings of the IEEE Conference on Advanced Video and Signal Based Surveillance*, Florida, 2003, pp. 351–358, ISBN 0-7695-1971-7.

82 S. Arulampalam, S. Maskell, N. Gordon, and T. Clapp. A tutorial on particle filters for on-line non-linear/non-Gaussian Bayesian tracking. *IEEE Transactions on Signal Processing*, 2002;50(2):174–188.

83 T.M. Rath and R. Manmatha. Features for word spotting in historical manuscripts. In *Proceedings of the 7th International Conference on Document Analysis and Recognition*, 2003, pp. 512–527, ISBN 0-7695-1960-1.

84 T. Oates, M.D. Schmill, and P.R. Cohen. A method for clustering the experiences of a mobile robot with human judgements. In *Proceedings of the 17th National Conference on Artificial Intelligence and 12th Conference on Innovative Applications of Artificial Intelligence*, AAAI Press, 2000, pp. 846–851.

85 N.T. Nguyen, H.H. Bui, S. Venkatesh, and G. West. Recognising and monitoring high-level behaviour in complex spatial environments. In *IEEE International Conference on Computer Vision and Pattern Recognition*, Wisconsin, 2003, pp. 1–6.

86 Y. Ivanov and A. Bobick. Recognition of visual activities and interaction by stochastic parsing. *IEEE Transactions of Pattern Recognition and Machine Intelligence*, 2000;22(8):852–872.

87 C.S. Regazzoni, V. Ramesh, and G.L. Foresti. Special issue on video communications, processing, and understanding for third generation surveillance systems. *Proceedings of the IEEE*, 2001;89(10):1355–1365.

88 S. Gong and T. Xiang. Recognition of group activities using dynamic probabilistic networks. In *Ninth IEEE International Conference on Computer Vision*, France, 2003, vol.2, pp. 742–750, ISBN 0-7695-1950-4.

89 H. Buxton. Generative models for learning and understanding scene activity. In *Proceedings of the 1st International Workshop on Generative Model-Based Vision*, Copenhagen, 2002, pp. 71–81.

90 Y. Ivanov, C. Stauffer, A. Bobick, and W.E.L. Grimson. Video surveillance of interactions. In *Second IEEE International Workshop on Visual Surveillance*, Colorado, 1999, pp. 82–91, ISBN 0-7695-0037-4.
91 M. Christensen and R. Alblas. V^2 – *Design Issues in Distributed Video Surveillance Systems*. Denmark, 2000, pp. 1–86.
92 X. Yuan, Z. Sun, Y. Varol, and G. Bebis. A distributed visual surveillance system. In *IEEE Conference on Advanced Video and Signal Based Surveillance*, Florida, 2003, pp. 199–205, ISBN 0-7695-1971-7.
93 L.M. Garcia and R.A. Grupen. Towards a real-time framework for visual monitoring tasks. In *Third IEEE International Workshop on Visual Surveillance*, Ireland, 2000, pp. 47–56, ISBN 0-7695-0698-4.
94 C.-H. Wu, J.D. Irwin, and F.F. Dai. Enabling multimedia applications for factory automation. *IEEE Transactions on Industrial Electronics*, 2001;48(5):913–919.
95 L. Almeida, P. Pedreiras, J. Alberto, and G. Fonseca. The FFT-CAN protocol: why and how. *IEEE Transactions on Industrial Electronics*, 2002;49(6):1189–1201.
96 M. Conti, L. Donatiello, and M. Furini. Design and analysis of RT-ring: a protocol for supporting real-time communications. *IEEE Transactions on Industrial Electronics*, 2002;49(6):1214–1226.
97 L.E. Jackson and G.N. Rouskas. Deterministic preemptive scheduling of real-time tasks. *Computer, IEEE*, 2002;35(5):72–79.
98 P.E. Rybski, S.A. Stoeter, M. Gini, D.F. Hougen, and N.P. Papanikolopoulos. Performance of a distributed robotic system using shared communications channels. *IEEE Transactions on Robotics and Automation*, 2002;18(5):713–727.
99 M. Valera and S.A. Velastin. An approach for designing a real-time intelligent distributed surveillance system. In *Proceedings of the IEE Workshop on Intelligent Distributed Surveillance Systems*, London, 2003, pp. 42–48, ISSN 0963-3308.
100 M. Greiffenhagen, D. Comaniciu, H. Niemann, and V. Ramesh. Design, analysis, and engineering of video monitoring systems: an approach and a case study. *Proceedings of the IEEE*, 2001;89(10):1498–1517.
101 M. Matijasevic, D. Gracanin, K.P. Valavanis, and I. Lovrek. A framework for multiuser distributed virtual environments. *IEEE Transactions on Systems, Man, and Cybernetics*, 2002;32(4):416–429.
102 P. Castelpietra, Y.-Q. Song, F.S. Lion, and M. Attia. Analysis and simulation methods for performance evaluation of a multiple networked embedded architecture. *IEEE Transactions on Industrial Electronics*, 2002;49(6):1251–1264.

Chapter 2

Monitoring practice: event detection and system design

M. S. Svensson, C. Heath and P. Luff

2.1 Introduction

Over the past decade, we have witnessed a widespread deployment of surveillance equipment throughout most major cities in Europe and North America. Perhaps most remarkable are the ways in which closed circuit television (CCTV) has become a pervasive resource for monitoring and managing behaviour in public places in urban environments. The deployment of CCTV and surveillance equipment has been accompanied by the development of a particular form of workplace, the surveillance or operations centre. For example, if you take central London, a range of organisations including the police, London Underground, Westminster Council, British Transport Police, Oxford Street Trade Association, Railtrack and the like have developed operation centres that enable the surveillance of public areas and provide resources for the management of problems and events. These centres of coordination are primarily responsible for deploying organisation, that is, providing staff, widely dispersed within the remote domain, with the resources to enable and in some cases engender the systematic coordination of activities. In other words, personnel in the operation centres transform information that is made available to a centralised domain and use that information to provide others, in particular mobile staff, with ways of seeing events and responding to problems that would otherwise remain unavailable.

Aside from the political and moral implications of the widespread deployment of CCTV equipment, it also generates severe practical problems for the operators of the surveillance equipment. It is recognised that the 'human operator' based in an operations centre or control room can only monitor or remain aware of a small part of the information that is made accessible through numerous cameras and that significant problems and events are often overlooked. In this light, there is a growing interest amongst operators, as well as engineers and researchers, in the design and deployment of image recognition systems that can analyse visual images and automatically identify events and inform operators. Researchers and developers in the field of computer vision and image processing have during the last thirty years developed technologies and computer systems for real-time video image processing and event recognition [1,2]. These image recognition capabilities build on advanced image processing techniques for analysing the temporal, structural and spatial elements of motion information in video images. These technologies are able to recognise stationary objects, crowds and densities of individuals, patterns and direction of movements and more recently particular actions and activities of individuals. These systems may be able to provide resources for supporting surveillance personnel as part of their day-to-day activities. These systems need not replace staff or necessarily automatically undertake some action based on what they detect. Rather, they could support individuals to carry out their everyday monitoring as one part of their organisational and collaborative work. In particular, they can provide resources for skilled staff to enhance their awareness of individuals, events and activities in a remote domain.

These practical concerns and technical developments reflect a long-standing interest in the social and cognitive sciences with awareness and monitoring and the practices and reasoning on which people rely to discriminate scenes and to identify and manage problems and difficulties that may arise (see, e.g. References 3–5). In human factors and ergonomics, for example, situation awareness has become increasingly relevant for system designers in the development and evaluation of operator interfaces and advanced computer systems in areas such as air-traffic management, nuclear power control and medical monitoring (e.g. References 6–8). In the community of computer-supported co-operative work (CSCW), there is also a growing interest in awareness and monitoring, both social and technical research. In particular, there is a body of research concerned with the development of context-aware systems and technologies that can automatically sense changes in another context and distribute event information to people concerned [9,10]. Developers and researchers involved in the design of awareness technologies in human factors and CSCW share many of the concerns of those currently developing image-processing systems, for example, which events and activities can and should be detected, how these should be made apparent to the users of the system and how this support can be made relevant (but not intrusive) in the users' everyday work. What is particularly interesting with the current development of image processing is that these technologies are being considered to support individuals in settings that have complex configurations of technology, that are complex spatial environments and that involve complex collaborative arrangements between staff. Despite the long-standing interest in awareness and monitoring, the design of these systems and technologies seems to present novel challenges for supporting monitoring tasks in such complex and dynamic work settings. This may suggest why the deployment of advanced surveillance systems has remained somewhat limited and why it remains unclear how to take advantage of advanced automation and processing functionality without making surveillance work expensive and difficult for the staff to manage.

In this chapter, we wish to show how detailed understandings of everyday surveillance work and organisational conduct may inform the design and development of image processing systems to enhance the awareness and monitoring of complex physical and behavioural environments. The setting in question is the operations rooms of complex interconnecting stations on London Underground and other rapid urban transport systems in Europe. The research presented in this chapter is part of a recent EU-funded project (IST DG VII) known as PRISMATICA, concerned with enhancing support for security management including the development of technologies for surveillance work in the public transport domain. We will consider one particular system currently developed to support real-time monitoring of complex environments – an integrated surveillance architecture that can process data from multiple cameras and 'automatically' identify particular events [11]. The chapter will begin by exploring the tacit knowledge and practices on which operators rely in identifying and managing events and then briefly reflect on the implications of these observations for the design of technical systems to support surveillance. By examining how operators in this domain oversee a complex organisational setting, and the resource on which they rely to identify and manage events, we are considering the everyday practices of monitoring as well as the practical development of the technologies to support it.

2.2   Setting the scene: operation rooms

Each major station on London Underground houses an area which is known as the operations room or 'ops room' for short (Figure 2.1). It is normally staffed by one of the station supervisors who are responsible for overseeing the daily operation of the station and developing a coordinated response to problems and emergencies. At any one time there may be up to 30 additional staff out and about on the station, mostly station staff who are responsible for platforms, the ticket barriers and even the main entrance gates to the station. The operations rooms are normally located in the main entrance foyer to the station and include a large window which overlooks passengers entering and leaving the station, the barriers, the top of the escalators, the ticket machines and the ticket office.

This panorama is enhanced by a series of, normally eight, monitors embedded within a console. These monitors provide access to images from more than 100 cameras that are located throughout the station: on platforms, in interconnecting passageways, in stairwells, over escalators, in foyers and the various entrances to the station and in some cases in areas surrounding the station itself. A number of these cameras, known as 'omni-scans', allow the operator to tilt and zoom and where necessary to focus on particular people or objects and focus in on specific activities. The monitors present single images and the systems allow operators to select and assemble any combination of views (cameras). An additional monitor provides traffic information of the running times and location of particular trains on specific lines. Communications equipment includes conventional telephones and direct lines to line control rooms, a fully duplex radio which allows all staff in the station to speak to

Figure 2.1   Victoria Station operations room

each other and hear both sides of all conversations and a public address system which allows the supervisor or operator to make announcements to an area within the station. Station control rooms are part of a broader network of control centres through which they receive information and to which they provide information. These include the line control rooms, the London Underground Network Control Centre and the British Transport Police Operations' Centre.

For ease, we will focus on a particular station, namely Victoria. Victoria is one of the busiest stations on London Underground and connects directly to the main over-land railway station. It handles approximately 120,000 passengers a day, a substantial number of whom are commuters from the dormitory towns of south-east England, but also a significant number of tourists, both those visiting local sites near the station and also passengers travelling to and from Gatwick, the second busiest airport in the UK. Like other major urban transport stations both in the UK and abroad, Victoria suffers a range of characteristic problems and difficulties – problems and difficulties that staff have to manage on a day to day basis. These include for example, severe overcrowding in the morning and evening rush hours when it is not unusual to have to temporarily queue passengers at the entrance gates and ticket barriers. They include other routine problems such as petty crimes – ticket touting, pickpocketing, unofficial busking, drug dealing, disorderly behaviour and the like. They include other problems such as late running, signal failures and passenger accidents, such as people falling on escalators, drunkenness and the like. They also include less routine but nonetheless 'normal' problems such as suspect packages, fires and 'one unders' (people committing suicide by jumping in front of the trains).

2.3   Revealing problems

Operators are highly selective in what they look for and when. Routine problems and difficulties typically arise in particular locations at certain times of the day, week or even year. Operators configure the CCTV system to enable them to see, discover and manage the problems and difficulties that typically arise in particular locations at certain times. For example, at Victoria Station severe overcrowding arises during the morning weekday rush hour on the Victoria Line north-bound platform, and in turn this affects certain escalators and the build-up of passengers at particular ticket barriers. The operator will select and maintain those views of the morning peak and switch them to the south-bound Victoria platform by the mid-afternoon in expectation of the afternoon rush. Later in the evening, ticket touts attempt to trade in the main foyer, and it is not unusual to select a number of views of both the foyer and main entrances in order to keep an eye on developments and intervene if necessary. Elsewhere, in particular in the late afternoon and evening, unofficial buskers and beggars will position themselves at the bottom of escalators and in certain interconnecting passageways and the supervisor will attempt to select views that allow them to see what may, if anything, be happening. As crowds build up, late in the evening, operators select views that enable them to keep track of events that might arise on particular platforms and require timely intervention.

The routine association between location, conduct and organisationally relevant problems provides a resource for selecting a combination of images at any one time but remains dependent upon the supervisor's abilities to spot, identify and manage problems that may arise. Operators discriminate scenes with regard to incongruent or problematic behaviour and to certain types of people both known and unknown. For example, people carrying objects, such as musical instruments that may be used for busking, may be followed as they pass through different areas of the station, the elderly or infirm, groups of young children or even people underdressed or over-dressed for the particular time of year. They note people carrying large or unusual packages, people trying to get cycles or push-chairs through the underground or individuals who are using sticks to walk. They note passengers who may dawdle on platforms after several trains have been through, who stand at the extreme end of a platform or who stand too close to the entrance of the tunnel. They note these not because they have some prurient interest in the curious and strange but because such appearances can have potential organisational significance for them and the running of the station. Such fleeting images can suggest problematic activity. For example, individuals carrying large amounts of baggage up flights of stairs often leave some behind at the bottom – which can in turn invoke sightings by other passengers of 'suspect packages'. Similarly, cycles and other awkward items can cause congestion in passageways and platforms. People hanging around platforms may be potential victims of suicides or accidents. Hence, 'keeping an eye' on potentially problematic passengers from the normal ebb and flow may help deal with troubles later.

Some of the practices for discriminating scenes draw on procedures for recog-nising incongruities within the web of 'normal appearances'. Operators for example will notice two people, even within a crowd, who walk noticeably close to each other as they approach a ticket barrier and discover 'doublers' – passengers who try and squeeze through the gate two at a time. They will notice people who are not looking where others may be looking, in particular those who are looking down at other passengers, and thereby spot pickpockets. They will notice people who pass objects to others, those who walk in the wrong direction along a passageway and those who congregate in the areas though which people normally walk. These and a host of other practices and procedures provide operators with the ability to see and discover potential problems in a world that would ordinarily appear unproblematic, a world that is already discriminated with regard to a series of interconnected scenes and possibilities.

Station supervisors and other control room personnel, then, do not passively monitor the displays waiting for 'something to happen' or just glance at the screens to gain a general awareness of the activities occurring within the station. They actively discriminate scenes and have resources that enable them to notice, for all practical purposes, what needs to be noticed, that is people and events that might disrupt the orderly flow of passengers through the station.

2.4   'Off-the-world'

The world available through surveillance equipment such as cameras and monitors is a limited fragmented world, disjointed scenes taken from particular viewpoints. Despite the substantial number of cameras at Victoria and many major interconnecting Underground stations, the complexity of the environments, constructed over many years, with irregularly shaped passageways, corridors and stairways, means that areas are not covered by cameras; even if they are pan-tilt cameras, the views are highly restricted. For example, it is often difficult to see the end of a platform where certain incidents are likely to occur, corners of a passageway or the area just beyond a stairwell or escalator.

Those interested in using stations for activities other than travel are well aware of the restricted views afforded by cameras and undertake certain activities in domains that are less accessible to operators. Particular areas of the station, at junctions, at the top of stairs and at locations down a corridor are popular not only because they are inaccessible to CCTV but because they are on the regular routes of passing passengers. Operators are well aware of the limitations of their views and use what they can see to determine what may be happening 'off-the-world' of the camera. For example, they will notice passengers swerving as they walk down a passageway and reason that there must be some obstruction. On closer inspection they may notice that certain passengers motion their hand as they swerve to avoid the obstacle, thereby inferring someone is taking money by begging or busking. Or for example, the supervisor will notice that passengers are joining an escalator at its base but leaving in dribs and drabs, finding that there must be some blockage in between: a fallen passenger or other form of obstacle. The views through the cameras and windows of the operation rooms therefore provide the supervisors with the ability to use what they can see to discover problems and events that occur elsewhere, problems and events that demand further analysis either by using a pan-zoom camera, flicking to other views, or by sending a member of staff to investigate.

Therefore, the use of CCTV, and the views that it affords, relies upon the supervisor's abilities to know what lies behind and beyond the image afforded through the camera. It provides access to a known world of organised spaces, objects and obstructions and of routine patterns of conduct and activity. In looking and connecting scenes, the supervisor relies upon his ability to infer, in situationally relevant ways, what there is and what could serve to explain the behaviour. Without a detailed understanding of the world beyond the image and its practical coherence, the images themeselves provide few resources with which to discriminate activities in situationally relevant and organisationally appropriate ways, identify events and implement practical solutions.

2.5   Combining views

We can begin to see that the detection and determination of particular problems and events may involve more than a single view or image. For example, operators may use the pan-tilt cameras to adopt a more suitable viewpoint to examine an event or may select adjoining camera views in order to make sense of what is happening. Consider for example, a seemingly simple problem that for long received attention from those interested in developing image recognition systems for urban applications, namely 'overcrowding'. One might imagine that for those interested in developing systems to automatically identify events, the density of people in a particular area could be subject, unproblematically, to detection. However, the operationally relevant distinction between 'crowding' and 'overcrowding' may not be as straightforward as one might imagine.

During the morning and evening weekday rush hours, areas of Victoria Station on London Underground, like any other rapid urban transport system, are frequently crowded. On occasions, this crowding is seen as demanding intervention. Consider a brief example. We join the action during the morning 'peak' in the operations room at Victoria. There are delays on the Circle and District Line and these are leading to a build-up of passengers on the platform. In the case at hand, the supervisor finishes a phone conversation and then glances at a series of monitors.

The supervisor turns and looks at the traffic monitor on his right (see Figure 2.2(a)), listing the trains on the Circle and District Line and their projected arrival times at the station. He then glances at the screen showing the west-bound platform (Figure 2.2(b)), the screen showing the foyer which feeds the Circle and District Line platforms (Figure 2.2(c)) and once again, glances back at the west-bound platform (Figure 2.2(d)). The supervisor then initiates a series of actions to reduce the number of passengers reaching the platform and in turn the number of passengers arriving at the ticket barriers. He requests passengers to be held at the ticket barriers and at the main entrance gates to the station. As the subsequent announcement reveals, his actions are designed to avoid overcrowding on the west-bound platform on the Circle and District Line.

The supervisor makes a practical assessment of overcrowding and implements a course of action to reduce the number of passengers arriving on the platform. The assessment is not based simply on seeing that the platform is crowded – it is crowded most of the time during the morning rush hour – but rather with regard to interweaving

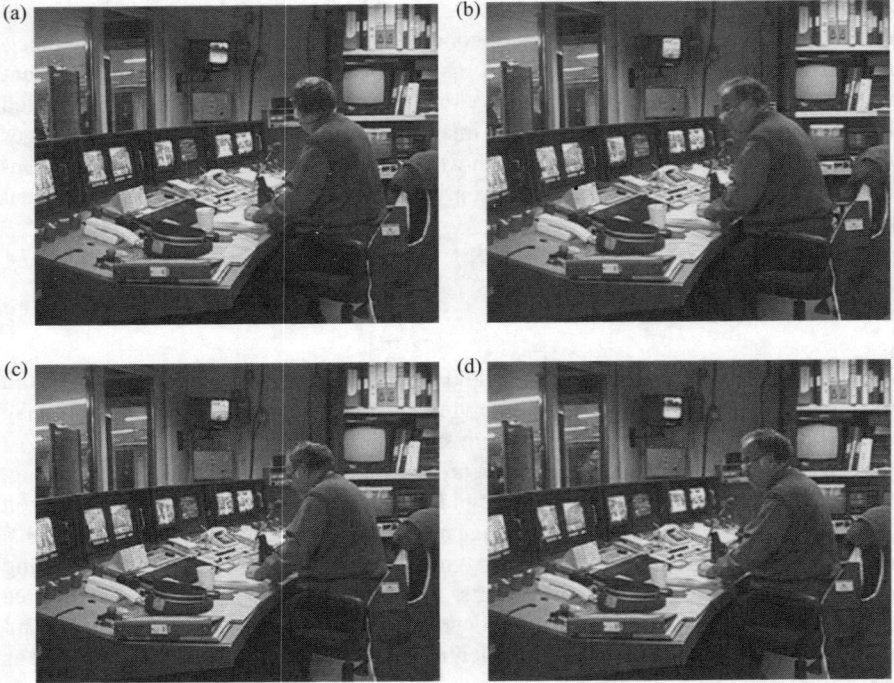

SS: Erm (1.2) Station Control for the West and er keep them outside
the station again please ...... Ladies and gentlemen we
apologise for keeping you outside the station ..... this is to
prevent overcrowding on our west-bound platform and in our ticket hall areas.

Figure 2.2   Fragment 1

distinct domains within the station – the foyer and the platform. The supervisor sees
what is happening within the foyer with regard to the number of passengers waiting on
the platform, just as seeing people crowded on the platform recasts the significance of
the number of people within the foyer. The supervisor configures an arrangement of
scenes from the CCTV monitors with which to assess the number of people within the
station and in particular to decide whether the station and the platforms in question
are becoming, or may become, overcrowded. The relevant scenes are configured
with regard to the routine patterns of navigation undertaken by passengers through
the station and in particular their movement from entrance gates, through the foyer,
to the platform. Whether the station or particular areas therefore are, or are not
overcrowded, or may become overcrowded, is not simply based upon the density of
passengers waiting on the platform.

The case at hand involves other considerations. Seeing or envisaging whether
the platform is overcrowded also depends upon the flow of traffic into the station.
Trains rapidly remove passengers from platforms so that in judging whether an area

is overcrowded, or will become overcrowded, depends in part whether passengers are likely to be on their way in a short time. The supervisor glances at the traffic monitor in order to see when the next train(s) will arrive, given what he knows about the build-up of passengers in various locations and whether the vehicles will have a significant impact on reducing the amount of passengers. The CCTV images therefore are examined with regard to the upcoming arrival of trains on particular platforms. It is not simply the density of passengers in particular locales, here and now, but rather the predictable pattern of events over the next few minutes.

In configuring and considering an assortment of potentially interrelated scenes, therefore, the supervisor is prospectively oriented. He considers the state of play in various potentially interdependent domains and how various events, such as passenger navigation and pace and train movement, will transform the scenes in question. In implementing station control by closing the ticket barriers and the entrance gates to the station, the supervisor is not simply responding to 'overcrowding' here and now; rather, he is envisaging what might happen and undertaking remedial action before problems actually arise. The simple automatic detection of an image that shows a crowded platform would be largely irrelevant and in many cases too late to deal with the problems as they actually arise. A single scene or image, without knowing what is happening elsewhere, or envisaging what is about to happen, does not provide the resources to enable supervisors to recognise overcrowding and initiate an organisationally relevant solution.

2.6 Tracking problems: coordinating activities

Views are combined in others ways in order to identify and manage problems. Events arise through time, and it is often necessary to progressively select a series of views in order to track and make sense of the problem. Many of the problems the supervisor deals with during the day involve utilisation of video technology to follow incidents, passengers and events through the complex layout of the station. One of the more common problems during most of the day in many stations, and in particular at Victoria Station, is the activities of certain individuals named ticket touts or fare dodgers. These individuals are known to be engaged in collecting used one-day travel cards left behind by passengers on their way from the station, which then can be sold to other passengers wishing to travel for less money. Unfortunately, when passengers no longer need their travel card at the end of the day they leave their tickets on the machines at the gate barriers or drop the tickets on the floor. Usually, the ticket touts remain outside the station at the top of the stairs on the way out to ask for tickets from passengers. Sometimes they enter the station to collect tickets from the floor or on the ticket machines at the barriers on the way up from the platforms.

Apart from huge losses of revenue for the stations, the activities of these individuals create an environment of insecurity and stress for many passengers. One way of dealing with these problems is to keep an eye on these individuals and coordinate actions with remote staff to escort them away from the station when spotted. One of the issues concerning the enhancement of an individual's awareness of the remote domain is to consider the ways in which image recognition analysis can be integrated with other technologies such as face recognition systems in order to enhance the tracking of an individual over a series of cameras and support the supervisors to coordinate their actions with other members of staff in the remote locations.

Consider the following example. It is evening at Victoria Station and the supervisor is switching through a selection of camera views when he notices a familiar face on the screen showing a camera image of the stairs leading down to the district ticket hall.

The supervisor notices a known offender walking down the stairs to the ticket hall (see Figure 2.3(a)–2a) where the passengers leave the station after travelling on the District and Circle lines. He switches to another camera and uses the pan-tilt-zoom camera, moving it up and to the left until the individual appears on the screen again at the bottom of the stairs (Figure 2.3(a)–2b). He follows the ticket tout as he progresses through the ticket hall towards the gate barriers (Figure 2.3(a)–2c). He then activates his radio and issues an instruction to the staff at the barriers. In this ticket hall, passengers usually leave their tickets on the machines when passing through the barriers on their way up. Seeing the ticket tout entering the station, the supervisor knows that the individual may potentially take the route by the machines, discretely collecting the tickets left on the machine. Thus, the supervisor does not only encourage the staff at the barriers at the 'way up' to make sure no tickets are left on the machines but to keep an eye on individuals walking towards the barriers appearing to have other intentions than travel.

The supervisor continues tracking the movement of the individual, watching him bend down to pick up tickets from the floor (2d, see Figure 2.4). The ticket tout suddenly takes another direction that goes beyond the boundaries of the camera, passing out of sight (2e). The supervisor pans the camera towards the long subway leading away from the ticket hall and then selects an alternative camera. He spots the individual again and by chance two police officers walking behind one of the station assistants (2f). Again, he delivers an instruction to one of the station assistants also spotted in the camera views.

The supervisor has configured the cameras and found an opportunity to provide the resources for his colleague to locate the individual. In his message to the station assistant, he uses the presence of the police officers to identify for his colleague the location of the ticket tout and the support at hand to escort the individual outside the station or to continue the chase. Unfortunately, the station assistant seems to miss the opportunity to spot the ticket tout, despite further attempts to identify him and his way through the station, up the long subway. A moment later, the supervisor switches the image on the screen to a camera displaying the long subway from the other direction and notices the individual again. As the long subway is leading towards the foyer where the operations room is located, the supervisor knows that he will soon be able to see the ticket tout appearing outside his own window. He shifts his gaze away from the screen and glances towards the scene of the foyer (see Figure 2.5) and sees the ticket tout speeding across the ticket hall outside the operations room and towards the station exit (2g). At this moment, he can also see that the two police officers are walking along the subway and following the same path as

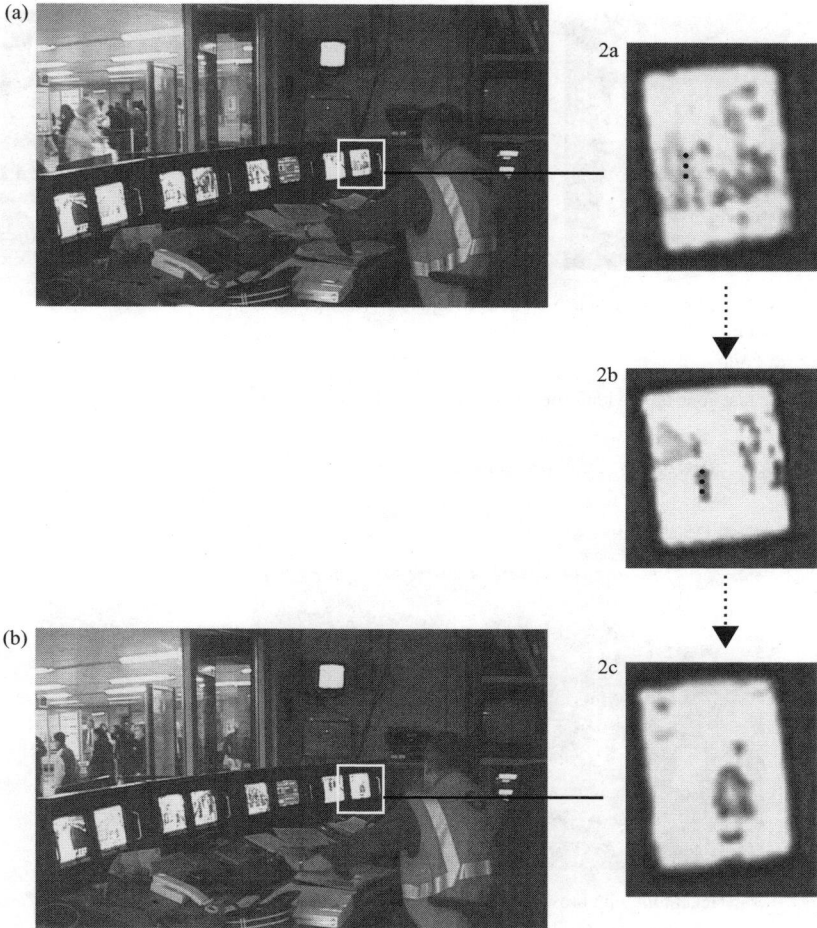

SS: District way out barrier staff we have a ticket tout approaching
the way up. Make sure there are no tickets left on the machines please.

Figure 2.3   Fragment 2: Transcript 1

the ticket tout. The supervisor takes the opportunity to use the PA to speak to the tout
directly.

In this announcement, he identifies for the ticket tout that he has been seen and is
still being watched. It is also an announcement for the general public and in particular
for the two police officers to identify the individual. The police officers appear outside
his window a moment later, briefly turning around towards the supervisor, which
nicely confirms for the supervisor that they have acknowledged the message and
perhaps have identified the individual.

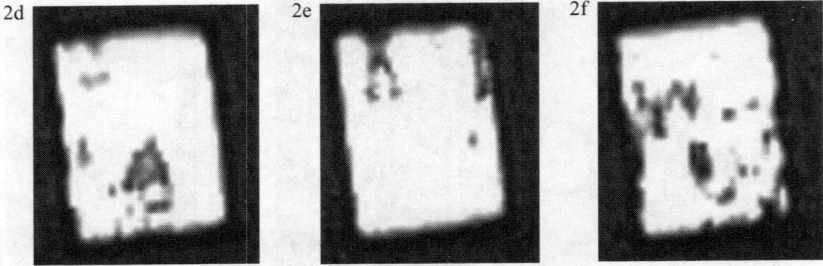

SS: Yeah listen Mr Dunn. You have two police officers behind you.
    Pull in that ticket tout now.

SA: Where is he?

SS: Just gone up the long subway (wearing) the brown jumper.

Figure 2.4   Fragment 2: Transcript 2

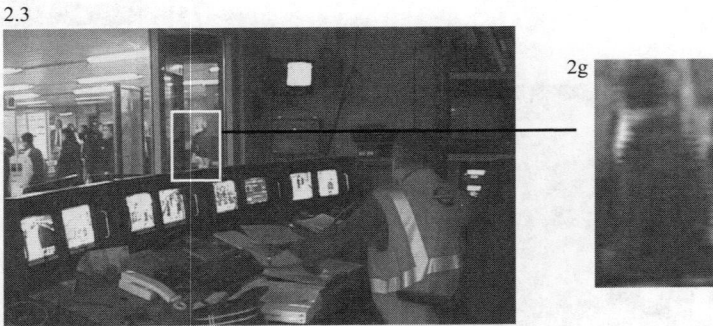

SS: Ticket tout with brown jumper, leave the station, you have
    been asked by staff to leave the station. Leave the station.
    Message to the ticket tout in brown, woollen jumper.

Figure 2.5   Fragment 2: Transcript 3

The identification of the event, both recognising the tout and seeing him in action, and its management, rests upon the supervisor's ability to combine a series of images, and of course his knowledge of how the actual scenes are connected by virtue of ordinary patterns of navigation. A single image alone, as in this example, the identification of an individual, is neither enough to identify the event nor to implement an appropriate course of action; the supervisor relies upon other emerging and highly contingent real-time analysis of the conduct of various participants. He follows the tout by subtly shifting camera angles and views, orienting according to conduct and prospective conduct of the various participants. He also has on camera one aspect of the crime: the collection of used tickets. The supervisors are not only trying to maintain awareness of the 'situation' in the station, their monitoring of the screens is tied to the local contingencies, the moment-to-moment demands and sequential import of their everyday activities. The very seeing by the surveillance of the scene, its meaning and intelligibility, derives from and is produced through the individual's orientation to the sequential relationship of particular events and particular activities. This interdependence between the perception of the scene, the particular events and the activities of personnel, forms the foundation of the surveillance and their highly organised discrimination of the views before them.

As the examples suggest in this chapter, the perception and identification of events is inextricably bound to the routine practices through which specific problems and events are managed. These practices routinely entail deploying collaboration, that is, encouraging staff and sometimes passengers to undertake specific courses of action. The deployment of collaboration relies upon the supervisor's abilities and using the resources he has available to make fellow staff, and passengers 'aware' of each other and events within the local milieu that might otherwise remain invisible and/or pass unnoticed. The CCTV and information systems provide the supervisors with the ability to reconfigure the ecology of an action or an event and to place it within a field of relevances that is unavailable to the participants (whether staff or passengers) within a particular scene. The coordination of activities and implementation of solutions to various problems involves in part providing other people in the remote domain with ways of perceiving how their own actions are related to events beyond their immediate perception and to providing resources through which they see and respond to each others' conduct and thereby engender co-operation and collaboration.

2.7   Deploying image processing systems: implications for design

The complex array of practice and reasoning that personnel rely upon in discovering and managing problems and difficulties provides a fundamental resource in the day-to-day operation of the service. The travel arrangements and working days of hundreds of thousands of passengers a week rely upon this seemingly delicate body of tacit procedure and convention. These practices provide personnel not only with ways in which they can selectively monitor the station, the conduct and interaction of passengers but with ways of envisaging problems and implementing solutions, in some cases even before those problems arise. Despite the robustness and reliability of these practices, it is recognised by staff and management and known by passengers that events frequently pass unnoticed and these events can lead to minor and some cases severe difficulties and threaten the safety of passengers. For those responsible for the day-to-day management of stations there are a number of familiar difficulties:

- At any one time, they have to select scenes from a small sub-selection of cameras. Problems and events frequently arise in domains that are not currently visible.
- The selection and monitoring of particular scenes is informed by a body of practice that is oriented towards detection of routine problems and events that arise

in specific areas at certain times. Problematic events and conduct that are not encompassed within these routine occurrences frequently pass unnoticed.

- Personnel engage in a range of activities whilst simultaneously 'monitoring' the various scenes, making announcements, writing reports, talking to staff, dealing with passenger problems, booking in maintenance staff, etc. Significant and problematic events can easily pass unnoticed, given the wide range of responsibilities and contingent demands on station supervisors.

One possible way of helping with this problem is to consider whether recent technical developments, in particular advanced image processing technologies, can enable more widespread surveillance and enhanced security management of public settings. It is recognised, amongst both staff and management, that the technology may provide relevant support for the detection, identification and management of events. There is, of course, in some circles, a naive belief that technology can solve the problem, even replace the human operator. But in general, it is recognised that the human operator should remain central to the daily operations of the surveillance cameras, and will need to determine the identification and relevance of events. It is critical that any system preserves the ability of personnel to detect and manage routine problems and difficulties whilst enhancing their abilities to oversee the station and maintain a reliable service. In the development of system support for surveillance, our empirical observations have informed the design of the technology, its assessment and a strategy to inform its deployment. In this section, we wish to discuss one or two issues that have derived from our field work and which we believe are critical to the successful development and practical application of the system.

2.7.1   Drawing upon local knowledge

Station personnel on London Underground, as in other urban transport systems, have wide-ranging practical knowledge of the routine behaviour of passengers and the time and location of problems and events. Even relatively rare difficulties, such as individuals falling onto the track, physical attacks and passenger accidents, are known to occur at certain times and in specific locations. Moreover, at any time of the day and night, staff are sensitive to the necessity to closely monitor particular domains and the limits of their ability to encompass all the locations in which certain difficulties might arise. In other words, the successful deployment of the system will draw upon, and support, the operator's familiarity with the station and its routine localised problems. The system will also have to draw upon their sensitivity to the occasioned circumstances, the here and now, within which relevant scrutiny of scenes and conduct is undertaken. Personnel require the ability to determine, for any occasion, the particular scenes that the system should monitor. They also require the ability to determine the range of events that the system will detect. For example, it might well be irrelevant to have the system monitoring for crowding on a particular platform when it is known, that at certain times of the day or night, the platform is routinely crowded. Alternatively, it may be critical to have the system monitor passenger conduct at a particular location on a platform, say late at night,

Image objects	Characteristics	Time/place	Events/problems
	Shape Static	Platforms, ticket halls	Suspect package
	Movements	Empty areas	Trespassing
	Stationary	Escalators Passageway	Busking, begging
	Stationary	Extreme end of platform	Suicides
	...	Station entries ticket hall	People carrying large objects
	Direction	Passageway Escalators	Wrong direction
	Speed		Running
	Patterns of positions and movements	Ticket machine areas	Pick-pockets Beggars
	Density Slow movements	Platform areas	Overcrowding
	Slow movements	Narrow spaces, passageways	Obstruction
...	...		...

Figure 2.6   Event characterisations

where the operator is committed to dealing with crowding in a different location and so forth.

Rather than processing large numbers of images for all kinds of different events, the local knowledge of the domain could help avoid unnecessary detection. The system should provide the possibility for the operators, on their own or together with a technician, of using a collection of 'image objects' such as physical objects, individuals and crowds (see Figure 2.6), recognised by various image processing modules, as the basic elements for configuring the actual detection of potential events and problems. These image objects would then be related to particular recognisable characteristics such as shape, stationarity, density, slow movement, direction and patterns of movements and so forth. As image events, these would have to be associated with camera views of particular places, or even specific areas visible in the image and the relevant time of the day or night. These image resources and event associations could then form the basis for specifying a collection of reasonably relevant event detections. For example, the operator could set a detection device to a camera overseeing a passageway or an escalator to monitor for the detection of individuals, or a small group of individuals, moving in the wrong direction, at all times of the day or night.

Allowing staff to easily implement associations between the discrete visual characteristics of the image with the spatial and temporal elements of the scene is critical to minimising the disturbance of a system informing operators about events they already know about or have already anticipated. In addition, the system should provide the ability to configure a selection of surveillance centre options encompassing different combinations of domains and system settings. These selections would allow the operators to switch between different settings depending on the changing demands during the day. The setting of the surveillance centre options would allow every network and individual station to configure their monitoring activities to their organisational geography and specific concerns. The collections of cameras, the events to be monitored, the detection thresholds and the time of day/week would allow the system to be useful despite the variability between networks and individual stations within networks.

Moreover, it is necessary that the system be able to provide system settings that not only allow operators to monitor domains where problems are likely to, or might, occur but also to set the system to detect conduct and events that forewarn that a particular problem may emerge or is currently emerging. Operators are sensitive to the prospective assessment of events and particular problems, that is, what ordinarily happens before a problem occurs. The skills and practices of staff in noticing and envisaging problems may suggest particular features and patterns of conduct and events that can inform the development of image recognition algorithms and enhance the detection of upcoming potential problems. For example, it might be relevant to have the system alert operators to large and oversized objects detected in the station as this may cause congestion and incidents, in particular during rush hours, at the top of escalators and on the way down to the platforms. The system therefore can allow operators to exploit their own skills and practices in detecting, identifying and envisaging problems and difficulties.

As we have also found in our observations, the practical and local knowledge of the staff may support the capability of the system to detect events that may only be available indirectly in the images. Despite the large number of cameras available to the operator, it is recognised that there are gaps in the video coverage. Problems that are relevant for the operator sometimes occur outside the scope of the camera images, or activities, that are known to be directly invisible, are deliberately undertaken beyond the world of the camera. Both operators and passengers are familiar with the location of these blind spots and the activities and events these spaces may conceal. It is therefore relevant to exploit the local knowledge of operators to set system detections that can detect such aspects of passengers' behaviour and conduct that may reveal problems and events only available for those actually in the scene.

Related to the configuration of event detections is the importance of making current detection settings and system operations apparent for the users at all time. Despite being locally tailorable, it is relevant to make apparent to the users of the system what actions the system is taking (and could take) and how the system has been configured. These considerations are of particular importance for systems taking automatic actions not initiated by the users themselves. An overview of the current system operations, easily available in the system interface, therefore, may provide useful information for envisaging and preparing for potential system identifications and the current performance with regard to their own and others' concerns (see example – Figure 2.7).

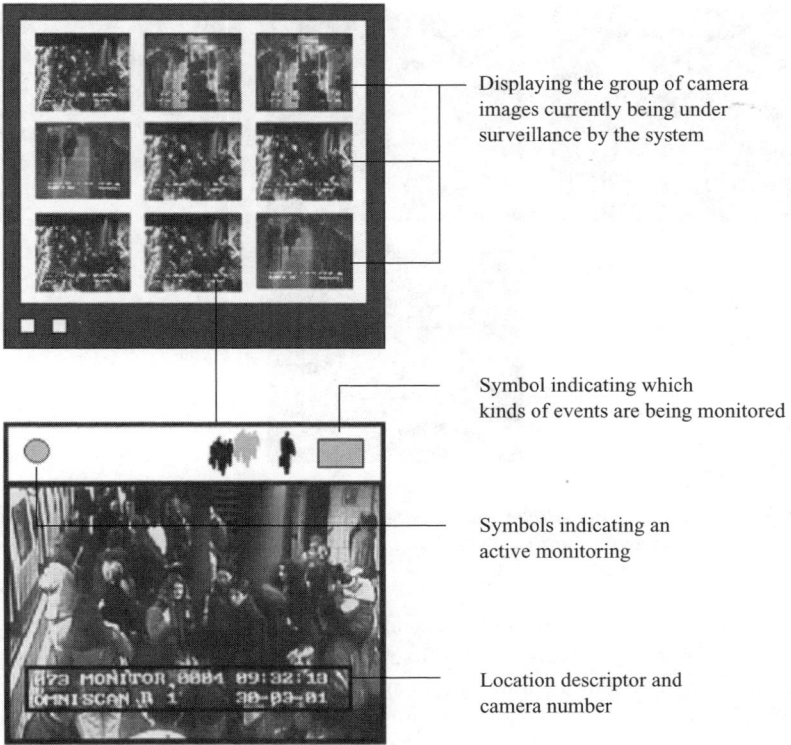

Displaying the group of camera images currently being under surveillance by the system

Symbol indicating which kinds of events are being monitored

Symbols indicating an active monitoring

Location descriptor and camera number

Figure 2.7   System overview – example interface

2.7.2   Combining images and data

Traditionally image recognition systems have been primarily concerned with detecting and identifying patterns from single images. To enable the system to provide relevant event detection for stations operations, it is necessary to combine images for analysis and data. We have seen from our studies how the detection and identification of many problems and events rest upon the operators' ability to combine images and data both simultaneously and sequentially. The most obvious example is overcrowding. An organisationally relevant identification of overcrowding rarely derives from a simple analysis of the number of people within a particular domain. Rather, it derives from an analysis of a combination of images selected from a number of interrelated scenes or locations. To provide the supervisor with relevant warnings, the system needs to combine an analysis of data from a number of relevant cameras, including for example, those on a platform, in a foyer and at an entrance gate (see Figure 2.8(a)). The detection of numerous other difficulties also requires analysis that combines related scenes. For example, detecting difficulties on an escalator requires the identification of a contrast between the density of movement of those joining

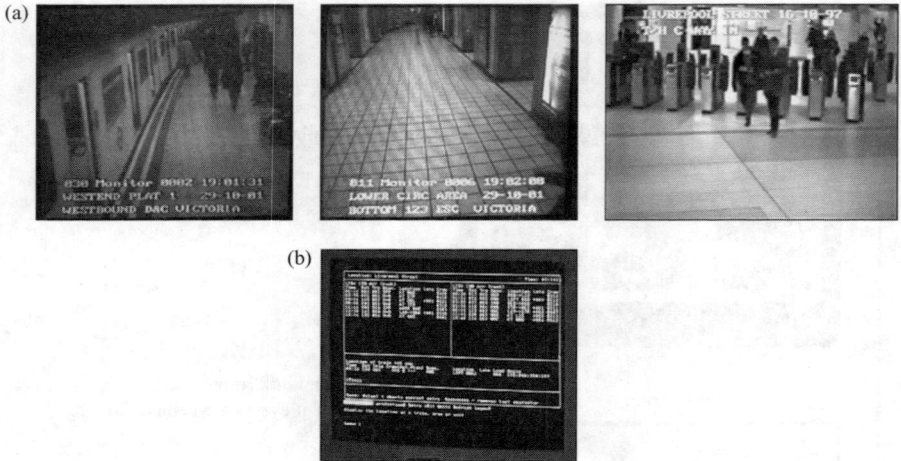

Figure 2.8   Combining information resources – platform areas, foyer to platforms, an entrance gate (a) and train traffic information (b)

and those leaving. Or for example, consider how problems in interconnecting passageways, in entrance foyers and the like may require a combination of images to detect a problem. Without combining images, images that are programmed to interlink relevant domains and domains that are interconnected through routine patterns of navigation and behaviour, the detection of many problems would prove to be either unreliable or impossible.

The accurate and relevant detection of problems can also require associated data that inform the analysis and identification of events. The most significant data that have to be taken into account for a number of problems concern traffic. The timing and in particular the arrival time of the next train, and trains immediately next, on a particular line is critical to the detection and management of problems, in particular crowding and overcrowding (see Figure 2.8(b)). The arrival times of trains provide a critical resource in prospectively envisaging where problems might or will emerge. At other times and for other events such information also proves of some importance, for example, when operators have to reorganise the movement of passengers when trains are being reversed at a station due to an incident further along the line or when evacuations are occurring and traffic is still passing through the station. Indeed, the timing and arrival of trains is critical to the relevant detection of various problems and their management by staff.

Other data are also relevant and we are currently exploring ways in which they can inform image analysis and the detection of problems. Consider for example, the ways in which information concerning broken escalators, gate closures, traffic flow on adjoining lines, arrival times of trains in mainline stations and major events such as football matches, marches, processions and the like can be critical to the analysis of visual images and the detection of particular problems.

Combining images with selected data to inform analysis and detection becomes still more complex when you consider the temporal emergence of an event. The system needs to select particular images through, ideally, a specified set of locations, over a period of for example, 3 minutes and develop an analysis drawing on related data, such as traffic information. In other cases, initial events require detection, perhaps using one or more cameras and associated data, but then the event itself requires tracking. For example, scuffles and fights routinely progress through different locations, and it is critical for staff to know at least the pattern of movement of the event. Or, for example, consider the importance of interconnecting an event in one area, such as an accident, with an event in another, such as an uneven flow of passengers from an escalator or stairwell. These complications raise important issues for the analysis of data from CCTV images and point to the difficulties in developing even basic systems that can enhance the work of supervisors and station staff. They are also critical to the police and others who require recordings of an event as it progresses in order to pursue and prosecute offenders and the like.

2.7.3   Displaying events: revealing problems

The detection and identification of events is only part of the problem. Station operation rooms are highly complex environments in which personnel, primarily the supervisor, have to oversee the domain, management events that arise, whilst engaging a range of more or less related tasks, dealing with maintenance crews, staff problems, passenger complaints and equipment failures. Alongside these tasks supervisors have to produce a range of associated documentation, reports on events and the like. It is important therefore that the system complements the detection and management of problems and events, whilst preserving the supervisor's ability to manage his routine tasks and responsibilities. How events are presented or displayed to the supervisor is critical to the ease with which they can be clarified, monitored, managed and, undoubtedly in some cases, ignored.

In this regard, a couple of important issues come into play at the outset. The system is unable to tell how busy a supervisor might be at any moment: it cannot tell whether the supervisor is already aware of the problem, and most critically, whether the system detection is reliable. In consequence, the system has to be carefully configured with regard to the certainty the system assigns to the detected events. It also provides a range of different alarms and notifications ranging from those requiring urgent attention to those that may indicate features of potential interest to the supervisor. Depending on the type of event and the expected detection rate, the system can provide options ranging from passive or peripheral indications, such as changing images on displays, to more obtrusive informings, such as alerts and alarms indicated in the system interface. Changing images on displays in the operations room may enable the operator to notice events detected by the system that may not be of direct relevance here and now or events that the operator, if noticed, can deal with when opportunities are given. We are also exploring whether ambient sound and other representation of potential events can provide opportunities to integrate image processing capabilities within actual working contexts without being overly disruptive.

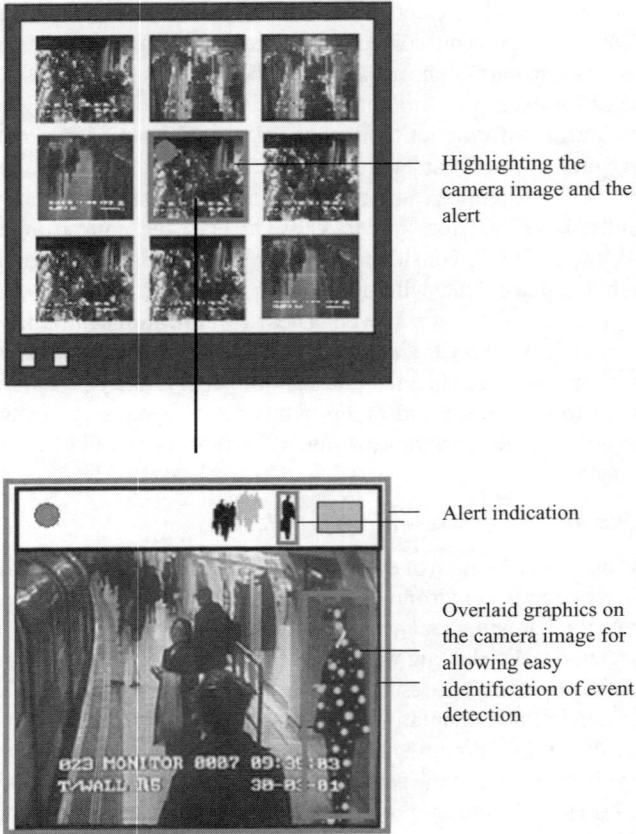

Figure 2.9   Highlighting and presenting event information

Providing an alert or warning is just one element of the problem. The supervisor requires relevant information concerning the event including for example, its location within a scene or scenes, its specific elements, its character and candidate identification or classification. The system therefore presents information so as to reflect, or better embody, the event as identified by the system and the resources it has used to detect it. The representation of the event can include an image or images coupled with overlaid graphics to demarcate the relevant features of the event coupled if relevant with text and even audio (see Figure 2.9). We are also currently exploring the potential benefits of combining a still image(s) with a recorded video sequence to support the operator in seeing events retrospectively. For example, this function may be relevant for the presentation of the detection of suspect packages, in which the video sequence can provide resources for confirming and identifying the relevance of the detection.

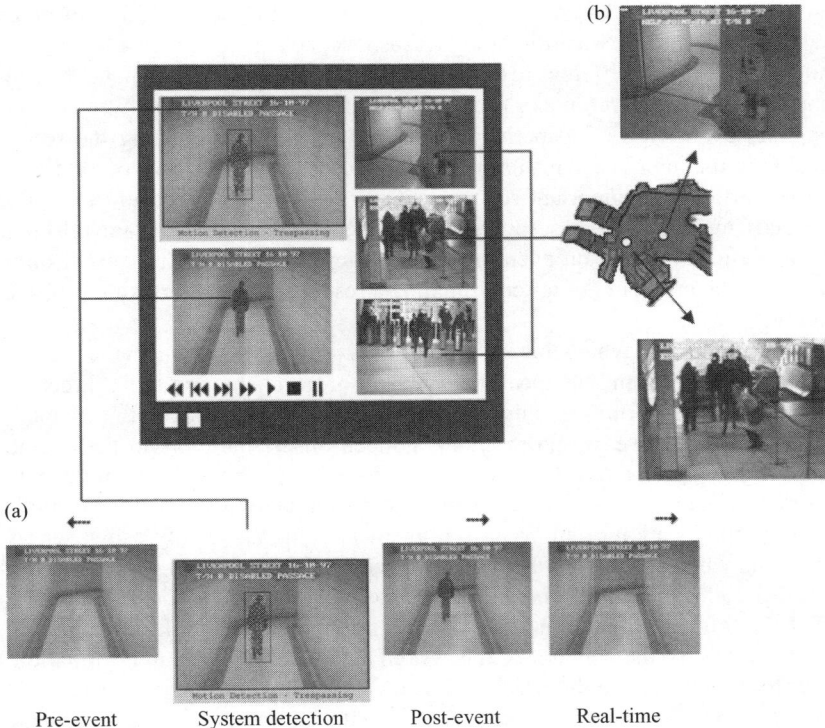

Figure 2.10   Video replays (a) and extended views of the local domain of the event (b)

This highlights a further aspect of the presentation of an event. Whilst a still image(s) and a recorded video sequence may illustrate the reason for the alert, it is important to have video sequence(s) to enable an operator not only to see what has happened but to see and determine what is currently happening. It is necessary therefore that the presentation of the event provides a video sequence, taken from one or more cameras, and in cases, enables continuous access to the relevant scenes as the event unfolds (see Figure 2.10(a)). It is particularly important that the system is able to display a combination of images, images that are relevant to the ways in which the operator oversees the local domain and the potential event. For example, the system presents additional camera images in the system interface that provide an extended view of a single event image such as the provision of images from nearby cameras along the known 'route' of particular individuals within the local domain (Figure 2.10(b)). It has to be stressed, however, that the system not only has to allow for the specification of resources with regard to the topology of the domain but, more importantly, with regard to the type and the nature of the particular event. The specification of a set of additional images and recordings based on camera location may undermine the ability to identify and manage a variety of events detected in the camera images. For example, the identification of an over-crowding event and the identification of a suspect package in the same camera image would require rather different resources for the verification and management of the events. The system therefore has to allow the staff to easily switch between topology-views and event-views and perhaps other options for manipulating the resources available in the interface. The ability to witness the relevant scenes of an unfolding event is particularly important for the operator as it provides resources to support other staff members, or passengers, in the local scene with relevant information and instructions. Seeing an event and its development can be critical not only for identifying the problem but for coordinating a response and undertaking prospective activities.

Another concern, which this issue highlights, is the ways in which the system can support mobile staff and the local management of events by providing direct access to information concerning events. As in the empirical examples in this chapter, we can see that many events and problems noticed on the monitors in the operations room are directly relevant for others in the domain. Even though many events are coordinated and managed from the centralised command, it is often local staff who are sent to the location of an event to implement solutions locally. However, as the operator may be engaged in other activities within the operations room or in contact with other colleagues, it may not be possible to provide a continuous update for local staff. Therefore, one of the ways the system can support the coordination and the local management is for the operator or the system automatically to send information and images to staff through mobile display devices. These images can be used as resources for seeing events that they are about to deal with on arrival or events occurring in other locations that may have an influence on their actions in other domains in which the problem may have emerged.

2.8   Summary and conclusion

The consideration of image processing systems in the surveillance domain pro-vides an interesting example of the challenges of supporting awareness in complex organisational settings. In this chapter we have examined the work of surveillance personnel in the urban transport domain and the everyday practices of monitoring and operating the service at individual stations. The character of work in the oper-ations room in London Underground stands in marked contrast to the conventional idea of surveillance found in some contemporary research in the social and cognitive sciences. Unlike the conventional idea of surveillance, their conduct rarely involves generalised 'monitoring' of the domain, rather CCTV and other information sources are vehicles through which they identify and manage a relatively circumscribed set of routine problems and events. To a large extent supervisors are not concerned with identifying particular individuals but rather with detecting conduct and events which might disrupt or are disrupting ordinary flow of passengers and traffic through the station. The routine problems and events which arise, and in particular the ordi-nary and conventional ways in which they manage those difficulties, inform the very ways in which they look at and use the images on the screen, their perception of the world through CCTV inseparable from their organisationally relevant and accountable practices for dealing with particular problems and events.

The design and deployment of image recognition systems to support work and coordination in the operation centres of London Underground raises some interesting technical problems as well as issues for the social and cognitive sciences. Even if we take a seemingly unambiguous problem such as overcrowding, we find no necessary correspondence between the density of passengers and its operational definition; indeed its organisational and practical characterisation has more to do with practical management of the free flow of passengers rather than the number of people walking or standing in the station. The picture becomes more complex still not just when we consider the recognition of overcrowding may depend on the inter-relationship of a series of 'interconnected' images and related information sources (such as traffic patterns) but also on a supervisor prospectively envisaging overcrowding rather than simply reacting when it arises. The perception and intelligibility of the scenes, or rather the images of those scenes, involves practical situated reasoning which relies upon the operator's ability to determine the scene with regard to an array of contextually relevant matters and concerns. It also depends upon the operator's practical concerns and the ways in which such actions and events are organisationally and routinely managed.

In this chapter we have discussed a number of design issues that are critical to the deployment of image recognition technology as support to event detection and everyday operational practices in the surveillance domain. It has been argued that the participants' familiarity with the local domain is a critical resource for setting up relevant and reliable event detections. It is important that the system be able to provide the users with flexible ways of configuring the event detections with regard to event characteristics in single images and their relevance with regard to particular places and the time of the day. It is also argued that many events need to be defined on the basis of a combination of camera views and other information resources such as train time schedules. Moreover, we have also shown that the presentation of an event is critical to the ways in which the operator not only can determine its relevance but can produce subsequent relevant actions and preserve the coordination of activities with remote staff and passengers. Information about events provided in the system has to be displayed and made available in such a way that it supports the practical activities of the operator and provides the resources for managing those activities and circumstances that other staff and passengers may be engaged in at other remote locations.

Acknowledgements

Part of the work described in this chapter was carried out in the project PRISMATICA, funded under the EC 5th Framework Programme DG VII. We are grateful for all discussions and joint work with partners in this project and all the generous comments which have contributed to this paper.

References

1 I.A. Essa. Computers seeing people. *AI Magazine*, 1999, 69–82.
2 R. Polana and R. Nelson. Detecting activities. In *Proceedings of Darpa Image Understanding Workshop*, Washington DC, April 1993, pp. 569–573.
3 M. Endlsey. Towards a theory of situation awareness in dynamic systems. *Human Factors*, 1995;37(1):32–64.
4 P. Luff, C. Heath, and M. Jirotka Surveying the Scene: technologies for every-day awareness and monitoring in control rooms. *Interacting with Computers*, 2000;13:193–228.
5 C. Sandom. Situation awareness through the interface: evaluating safety in safety-critical control systems. In *Proceedings of IEE People in Control: International Conference on Human Interfaces in Control rooms, Cockpits and Command Centres*, Conference Publication No. 463, University of Bath, June 21–23, pp. 207–212, 1999.
6 C.J. Arback, N. Swartz, and G. Kuperman. Evaluating the panoramic cockpit control and display systems. In *Proceedings of the Fourth International Symposium on Aviation Psychology*, 1987. Columbus, OH, The Ohio State University, pp. 30–36.
7 D.G. Jones. Reducing awareness errors in air traffic control. In *Proceedings of the Human Factors and Ergonomics Society 41st Annual Meeting*, Santa Monica, CA, Human Factors and Ergonomics Society, 1997, pp. 230–233.
8 W. McDonald. Train controllers, interface design and mental workload. In People in Control: Human Factors in Control Room Design, J. Noyes and M. Bransby, Eds. The Institution of Electrical Engineers, London, 2001, pp. 239–258.
9 L. Fuchs, U. Pankoke-Babatz, and W. Prinz. Supporting cooperative aware-ness with local event mechanisms: the group desk system. In *Proceedings of the Fourth European Conference on Computer-Supported Cooperative Work*, September 10–14, Stockholm, Sweden, pp. 247–262.
10 W. Prinz. NESSIE: an awareness environment for cooperative settings. In ECSCW'99: Sixth Conference on Computer Supported Cooperative Work, Copenhagen, S. Bödker, M. Kyng, and K. Schmidt, Eds. Kluwer Academic Publishers, 1999, pp. 391–410.
11 S.A. Velastin, J.H. Yin, M.A. Vincencio-Silva, A.C. Davies, R.E. Allsop, and A. Penn. Automated measurement of crowd density and motion using image processing. In *Proceedings of 7th International Conference on Road Traffic Monitoring and Control*, London, April 26–28 1994, pp. 127–132.

Chapter 3

A distributed database for effective management and evaluation of CCTV systems

J. Black, T. Ellis and D. Makris

3.1 Introduction

Image surveillance and monitoring is an area being actively investigated by the machine vision research community. With several government agencies investing significant funds into closed circuit television (CCTV) technology, methods are required to simplify the management of the enormous volume of information generated by these systems. CCTV technology has become commonplace in society to combat anti-social behaviour and reduce other crime. With the increase in processor speeds and reduced hardware costs it has become feasible to deploy large networks of CCTV cameras to monitor surveillance regions. However, even with these technological advances there is still the problem of how information in such a surveillance network can be effectively managed. CCTV networks are normally monitored by a number of human operators located in a control room containing a bank of screens streaming live video from each camera.

This chapter describes a system for visual surveillance for outdoor environments using an intelligent multi-camera network. Each intelligent camera uses robust techniques for detecting and tracking moving objects. The system architecture supports the real-time capture and storage of object track information into a surveillance database. The tracking data stored in the surveillance database is analysed in order to learn semantic scene models, which describe entry zones, exit zones, links between cameras, and the major routes in each camera view. These models provide a robust framework for coordinating the tracking of objects between overlapping and non-overlapping cameras, and recording the activity of objects detected by the system. The database supports the operational and reporting requirements of the surveillance application and is a core component of the quantitative performance evaluation video-tracking framework.

3.2   Background

3.2.1   *Multi-view tracking systems*

There have recently been several successful real-time implementations of intelligent multi-camera surveillance networks to robustly track object activity in both indoor and outdoor environments [1–6]. Cai and Aggarwal presented an extensive distributed surveillance framework for tracking people in indoor video sequences [1]. Appearance and geometric cues were used for object tracking. Tracking was coordinated between partially overlapping views by applying epipole line analysis for object matching. Chang and Gong created a multi-view tracking system for cooperative tracking between two indoor cameras [2]. A Bayesian framework was employed for combining geometric and recognition-based modalities for tracking in single and multiple camera views.

Outdoor image surveillance presents a different set of challenges, contending with greater variability in the lighting conditions and the vagaries of weather. The Video Surveillance and Monitoring (VSAM) project developed a system for continuous 24 h monitoring [3]. The system made use of model-based geo-location to coordinate tracking between adjacent camera views. In Reference 6 the ground plane constraint was used to fit a sparse set of object trajectories to a planar model. A robust image alignment technique could then be applied in order to align the ground plane between overlapping views. A test bed infrastructure has been demonstrated for tracking platoons of vehicles through multiple camera sites [5]. The system made use of two active cameras and one omni-directional camera overlooking a traffic scene. Each camera was connected to a gigabit Ethernet network to facilitate the transfer of full size video streams for remote access. More recently the KNIGHT system [4] has been presented for real-time tracking in outdoor surveillance. The system can track objects between overlapping and non-overlapping camera views by using spatial temporal and appearance cues.

3.2.2   *Video annotation and event detection*

One application of a continuous 24 h surveillance system is that of event detection and recall. The general approach to solving this problem is to employ probabilistic frameworks in order to handle the uncertainty of the data that is used to determine if a particular event has occurred. A combination of both Bayesian classification and hidden Markov models (HMMs) was used in the VIGILANT project for object and behavioural classification [7]. The Bayesian classifier was used for identification of object types, based on the object velocity and bounding box aspect ratio. An HMM was used to perform behavioural analysis to classify object entry and exit events. By combining the object classification probability with the behavioural model, the VIGILANT system achieved substantial improvement in both object classification and event recognition, when compared to using the object classification or behavioural models individually. The VIGILANT system also made use of a database to allow an untrained operator to search for various types of object interactions in a car park scene.

Once the models for object classification and behavioural analysis have been determined, it is possible to automatically annotate video data. In Reference 8 a generic framework for behavioural analysis is demonstrated that provides a link between low-level image data and symbolic scenarios in a systematic way. This was achieved by using three layers of abstraction: image features, mobile object properties and scenarios. In Reference 9 a framework has been developed for video event detection and mining that uses a combination of rule-based models and HMMs to model different types of events in a video sequence. Their framework provides a set of tools for video analysis and content extraction by querying a database. One problem associated with standard HMMs is that in order to model temporally extended events it is necessary to increase the number of states in the model. This increases the complexity and the time required to train the model.

This problem has been addressed by modelling temporally extended activities and object interactions using a probabilistic syntactic approach between multiple agents [10]. The recognition problem is sub-divided into two levels. The lower level can be performed using an HMM for proposing candidate detections of low-level temporal features. The higher level takes input from the low-level event detection for a stochastic context-free parser. The grammar and parser can provide a longer range of temporal constraints, incorporate prior knowledge about the temporal events given the domain and resolve ambiguities of uncertain low-level detection.

The use of database technology for surveillance applications is not completely new. The 'Spot' prototype system is an information access system that can answer interesting questions about video surveillance footage [11]. The system supports various activity queries by integrating a motion tracking algorithm and a natural language system. The generalised framework supports event recognition, querying using a natural language, event summarisation and event monitoring. In Reference 12 a collection of distributed databases was used for networked incident management of highway traffic. A semantic event/activity database was used to recognise various types of vehicle traffic events. The key distinction between these systems and our approach is that an MPEG-4-like strategy is used to encode the underlying video [13], and semantic scene information is automatically learned using a set of offline processes [14–18].

The Kingston University Experimental Surveillance system (KUES) comprises several components:

- Network of intelligent camera units (ICUs).
- Surveillance database.
- Off-line calibration and learning module.
- Multi-view tracking server and scheduler (MTS).
- Video summary and annotation module (VSA).
- Query human computer interface (QHCI).

The system components are implemented on separate PC workstations, in order to distribute the data processing load. In addition, each system component is connected to a 100 Mb/s Ethernet network to allow the exchange of data, as shown in Figure 3.1.

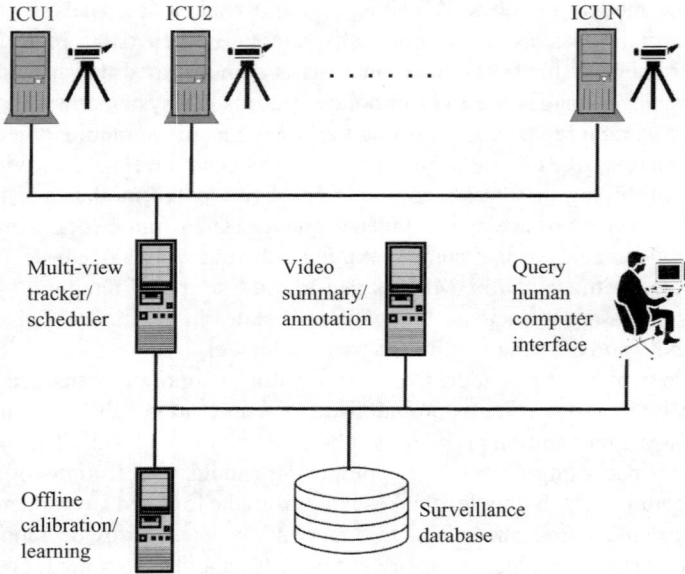

Figure 3.1   System architecture of KUES

The main functionality of each ICU is to robustly detect motion and track moving objects in 2D.

The tracking data generated by the ICU network are stored in the surveillance database. The system uses the PostgreSQL database engine, which supports geometric primitives and the storage of binary data. The stored data can be efficiently retrieved and used to reconstruct and replay the tracking video to a human operator. Transferring raw video from each ICU to the surveillance database would require a large network bandwidth, so only the pixels associated with each detected foreground object are transmitted over the network. This MPEG-4-like encoding strategy [13] results in considerable savings in terms of the load on the Ethernet network and storage requirements.

The tracking data stored in the surveillance database are analysed to learn semantic models of the scene [15]. The models include information that describes entry zones, exit zones, stop zones, major routes and the camera topology. The semantic models are stored in the surveillance database, so they can be accessed by the various system components. The models provide two key functions for the surveillance system.

First, the camera topology defines a set of calibration models that are used by the MTS to integrate object track information generated by each ICU. The model identifies overlapping and non-overlapping views that are spatially adjacent in the surveillance region. In the former case homography relations are utilised for corresponding features between pairs of overlapping views [19]. In the latter case the topology defines the links between entry and exit zones between the non-overlapping camera views. This model allows the MTS to reason where and when an object should reappear, having left one camera view at a specific exit zone.

Second, the semantic scene models provide a method for annotating the activity of each tracked object. It is possible to compactly express an object's motion history in terms of the major routes identified in each camera view. These data can be generated online and stored in the surveillance database. The metadata is a high-level representation of the tracking data. This enables a human operator to execute spatial–temporal activity queries with faster response times than would be possible if using the raw tracking data.

3.2.3 Scene modelling

We wish to be able to identify the relationship between activity (moving objects) in the camera's field of view and features of the scene. These features are associated with the visual motion events, for example, where objects may appear (or disappear) from view, either permanently or temporarily. Figure 3.2 illustrates two representations for the spatial information, showing the topographical and topological relationships between a specific set of activity-related regions – entry and exit zones (labelled A, C, E, G and H), paths (AB, CB, BD, DE, DF, FG and FH), path junctions (B, D and F) and stopping zones (I and J).

We can categorise these activity-related regions into two types. The first are fixed with respect to the scene and are linked to genuine scene features such as the pavement or road, where we expect to locate objects that are moving in well-regulated ways, or at the doorway into a building. The second are view-dependent, associated with the particular viewpoint of a camera, and may change if the camera is moved.

Figure 3.3 indicates a number of these activity zones that can be identified by observing the scene over long periods of time (typically hours or days). Entry/exit zones 1 and 3 are view-dependent; zone 2 is a building entrance and hence scene-related. The stop regions are examples of locations where people stop, in this case, sitting on the wall and the flag plinth.

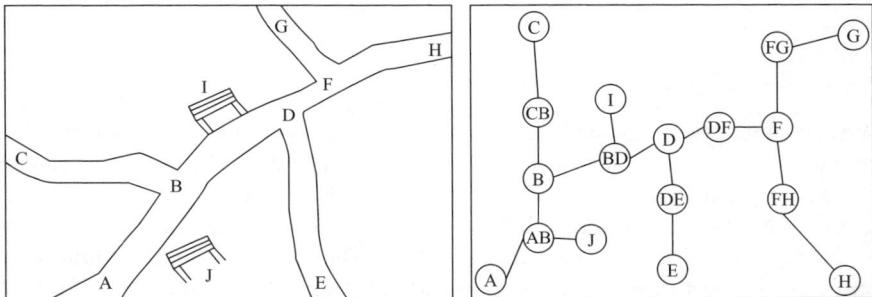

Figure 3.2 *(a) Topographical map showing the spatial relationships between activity-related scene features. (b) Topological graph of same scene*

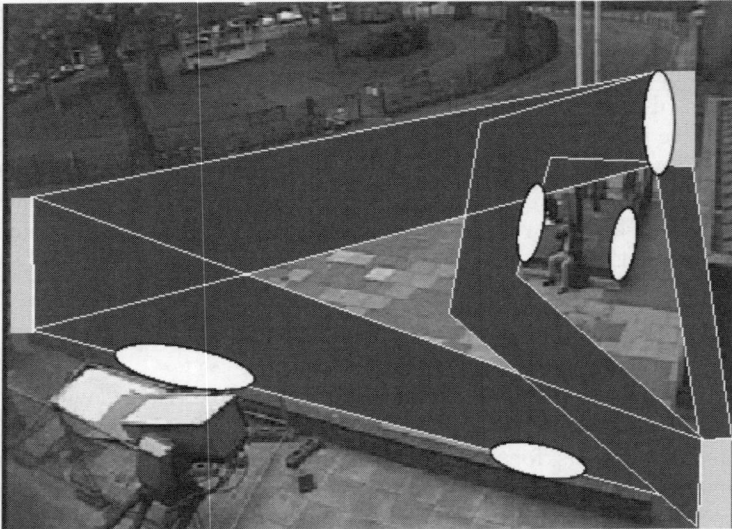

Figure 3.3   Manually constructed region map locating different motion-related activities in a camera field of view. Entry and exit zones are indicated by the light grey rectangles, paths are shown by the dark grey polygons and stop zones by the white ellipses

3.3   Off-line learning

The system performs a training phase to learn information about the scene by post-analysis of the 2D trajectory data stored in the surveillance database, as discussed in Section 3.1. The information learned includes semantic scene models and details of the camera topology. These are both automatically learned without supervision and stored in the surveillance database to support the functionality of the MTS, VSA and QHCI system components.

3.3.1   Semantic scene models

The semantic scene models define regions of activity in each camera view. In Figure 3.4 the entry zones, exit zones and routes identified for one of the camera views are shown. The entry zones are represented by black ellipses, while the exit zones are represented by white ellipses. Each route is represented by a sequence of nodes, where the black points represent the main axis of the route and the white points define the envelope of the route. Routes 1 and 2 represent lanes of vehicle traffic in the scene. It can be observed that the entry and exit zones are consistent with driving on the left-hand side of the road in the UK. The third route represents the bi-directional flows of pedestrian traffic along the pavement.

Figure 3.4   Example of semantic models stored in the surveillance database

3.3.2   Camera topology

The major entry and exit zones in the semantic scene models are used to derive the camera topology. The links between cameras are automatically learned by the temporal correlation of object entry and exit events that occur in each view [18]. This approach can be used to calibrate a large network of cameras without supervision, which is a critical requirement for a surveillance system. The camera topology is visually depicted in Figure 3.5, where the major entry and exit zones are plotted for each camera, along with the links identified between each camera view.

3.4   Database design

Multi-camera surveillance systems can accumulate vast quantities of data over a short period of time when running continuously. We address several issues associated with data management, which include the following: How can object track data be stored in real-time in a surveillance database? How can we construct different data models to capture multiple levels of abstraction of the low level tracking data, in order to represent the semantic regions in the surveillance scene? How can each of these data models support high-level video annotation and event detection for visual surveillance applications? How can we efficiently access the data?

3.4.1   Data abstraction and representation

The key design consideration for the surveillance database was that it should be possible to support a range of low-level and high-level queries. At the lowest level it

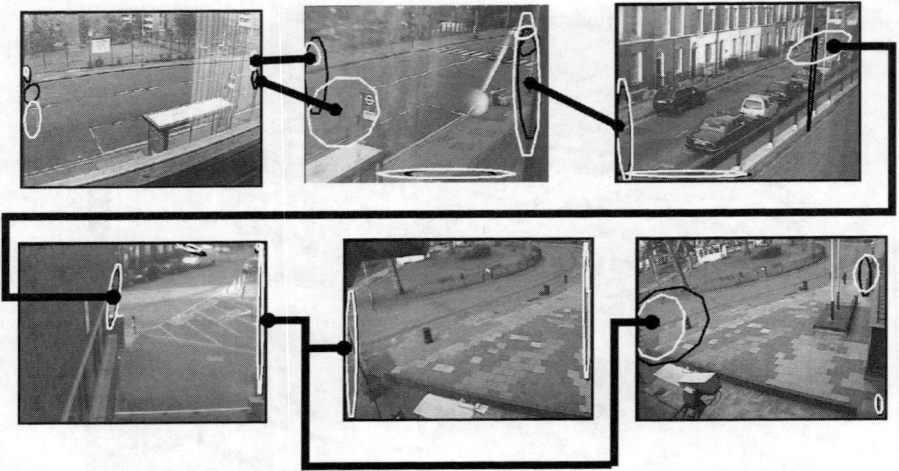

Figure 3.5   The camera topology learnt for a network of six cameras

is necessary to access the raw video data in order to observe some object activity recorded by the system. At the highest level a user would execute database queries to identify various types of object activity observed by the system. In addition to supporting low-level and high-level queries the database also contains information about each camera in the CCTV system, which makes it possible to define a schedule of when each camera should be actively recording tracking information. This functionally enables the seamless integration or removal of cameras from the CCTV system without affecting the continuous operation. This is an important requirement, since the CCTV network of cameras will undergo some form of maintenance during its operational lifetime. The surveillance database comprises several layers of abstraction:

- Image framelet layer.
- Object motion layer.
- Semantic description layer.
- Operational support layer.

This first three layers in the hierarchy support the requirements for real-time capture and storage of detected moving objects at the lowest level, to the online query of activity analysis at the highest level [20]. Computer vision algorithms are employed to automatically acquire the information at each level of abstraction. The fourth layer supports the operational requirements of the CCTV system, which includes storing details of each ICU along with its start-up parameters and mode of operation. The physical database is implemented using PostgreSQL running on a Linux server. PostgreSQL provides support for storing each detected object in the database. This provides an efficient mechanism for real-time storage of each object detected by the surveillance system. In addition to providing fast indexing and retrieval of data the

Figure 3.6   Example of objects stored in the image framelet layer

surveillance database can be customised to offer remote access via a graphical user interface and also log queries submitted by each user.

3.4.1.1   Image framelet layer

The image framelet layer is the lowest level of representation of the raw pixels identified as a moving object by each camera in the surveillance network. Each camera view is fixed and background subtraction is employed to detect moving objects of interest [21]. The raw image pixels identified as foreground objects are transmitted via a TCP/IP socket connection to the surveillance database for storage This MPEG-4-like [13] coding strategy enables considerable savings in disk space and allows efficient management of the video data. Typically, 24 h of video data from six cameras can be condensed into only a few gigabytes of data. This compares to an uncompressed volume of approximately 4 terabytes for one day of video data in the current format we are using, representing a compression ratio of more than 1000 : 1.

In Figure 3.6 an example is shown of some objects stored in the image framelet layer. The images show the motion of two pedestrians as they move through the field of view of the camera. Information stored in the image framelet layer can be used to reconstruct the video sequence by plotting the framelets onto a background image. We have developed a software suite that uses this strategy for video playback and review and to construct pseudo-synthetic sequences for performance analysis of tracking algorithms [22].

An entry in the image framelet layer is created for each object detected by the system. The raw image pixels associated with each detected object are stored internally in the database. The PostgreSQL database compresses the framelet data, which has the benefit of conserving disk space.

3.4.1.2   Object motion layer

The object motion layer is the second level in the hierarchy of abstraction. Each intelligent camera in the surveillance network employs a robust 2D tracking algorithm to record an object's movement within the field of view of each camera [23].

Features are extracted from each object including the bounding box, normalised colour components, object centroid and the object pixel velocity. Information is integrated between cameras in the surveillance network by employing a 3D multi-view object tracker [19] which tracks objects between partially overlapping and non-overlapping camera views separated by a short spatial distance. Objects in overlapping views are matched using the ground plane constraint. A first order 3D Kalman filter is used to track the location and dynamic properties of each moving object. When an object moves between a pair of non-overlapping views we treat this scenario as a medium-term static occlusion and use the prediction of the 3D Kalman filter to preserve the object identity when/if it reappears in the field of view of the adjacent camera.

The 2D and 3D object tracking results are stored in the object motion layer of the surveillance database. The object motion layer can be accessed to execute off-line-learning processes that can augment the object tracking process. For example, we use a set of 2D object trajectories to automatically recover the homography relations between each pair of overlapping cameras [19]. The multi-view object tracker robustly matches objects between overlapping views by using these homography relations. The object motion and image framelet layer can also be combined in order to review the quality of the object tracking in both 2D and 3D.

In Figure 3.7 results from both the 2D tracking and multi-view object tracker are illustrated. The six images represent the viewpoints of each camera in the surveillance network. Cameras 1 and 2, 3 and 4, and 5 and 6 have partially overlapping fields of view. It can be observed that the multi-view tracker has assigned the same identity to each object. Figure 3.8 shows the field of view of each camera plotted onto a common ground plane generated from a landmark-based camera calibration. 3D motion trajectories are also plotted on this map in order to allow the object activity to be visualised over the entire surveillance region.

3.4.1.3   Semantic description layer

The object motion layer provides input to a machine-learning algorithm that automatically learns a semantic scene model, which contains both spatial and probabilistic information. Regions of activity can be labelled in each camera view, for example, entry/exit zones, paths, routes and junctions, as was discussed in Section 3.3. These models can also be projected on the ground plane as is illustrated in Figure 3.9. These paths were constructed by using 3D object trajectories stored in the object motion layer. The grey lines represent the hidden paths between cameras. These are automatically defined by linking entry and exit regions between adjacent non-overlapping camera views. These semantic models enable high-level queries to be submitted to the database in order to detect various types of object activity. For example, we can generate spatial queries to identify any objects that have followed a specific path between an entry and exit zone in the scene model. This allows any object trajectory to be compactly expressed in terms of routes and paths stored in the semantic description layer.

Each entry and exit zone is approximated by an ellipse that represents the covariance of the region. Using this internal representation in the database simplifies spatial

Figure 3.7 *Camera network on university campus showing six cameras distributed around the building, numbered 1–6 from top left to bottom right, raster-scan fashion*

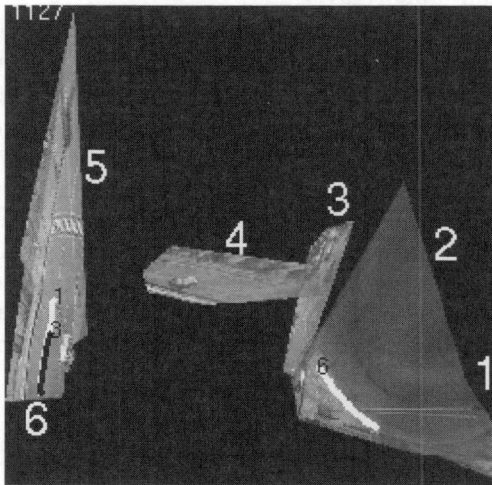

Figure 3.8 *Re-projection of the camera views from Figure 3.7 onto a common ground plane, showing tracked objects trajectories plotted into the views (white and black trails)*

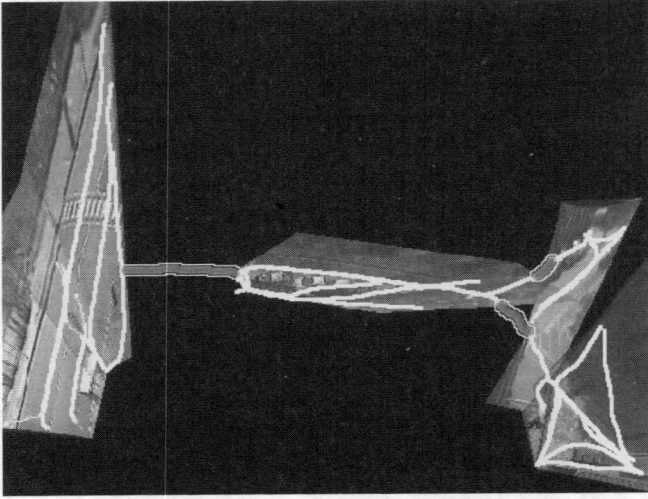

Figure 3.9   Re-projection of routes onto ground plane

queries to determine when an object enters an entry or exit zone. A polygonal representation is used to approximate the envelope of each route and route node, which reduces the complexity of the queries required for online route classification that will be demonstrated in the next section.

3.4.1.4   Operational support layer

The surveillance database incorporates functionality to support the continuous operation of the CCTV system. A scheduler accesses the database to determine when each ICU should be active and the type of data that should be generated. For example, if an intelligent camera is connected to a very low bandwidth network it is possible to suppress the transmission of framelets in order to reduce the required network traffic. The start-up parameters of each camera can easily be modified, which enables an ICU to execute different motion detection algorithms depending on the time of operation (night/day) or weather conditions. During the continuous operation of a CCTV network new cameras will be added or have to undergo routine maintenance. The QHCI component of the system architecture shown in Figure 3.1 provides a mechanism for security personnel to seamlessly manage the operation of the cameras connected to the CCTV system.

3.5   Video summaries and annotation

The surveillance system generates a large volume of tracking data during long periods of continuous monitoring. The surveillance database typically requires between 5 and 10 gigabytes for storing 24 h of tracking data. The vast majority of the space is consumed by the image data of each detected foreground object. A general task of a human operator is to review certain types of activity. For example, a request can be made to replay video from all instances where an object moved within a restricted area during a specific time interval. Even with appropriate indices the tracking data are not suitable for these types of queries, since the response times are of the order of minutes, which would not be acceptable to the operator if many similar queries had to be run. We resolve this problem by generating metadata that describes the tracking data in terms of the routes contained in the semantic scene models. Each tracked object is summarised in terms of its entry location, entry time, exit location, exit time and appearance information. To support spatio-temporal queries the complete object trajectory is stored as a path database geometric primitive. This storage format simplifies classification of an object's path in terms of the route models. Once the routes taken by the object have been determined, its complete motion history is expressed in terms of the nodes along each route. Metadata is generated that describes the entry time and location along the route, and the exit time and location along the route. When an object traverses several routes, multiple entries are generated. The data flow of the metadata generation process is shown in Figure 3.10. The operator can use the query interface to execute spatial–temporal queries on the metadata with faster response times (of the order in seconds). The tracking data can be accessed

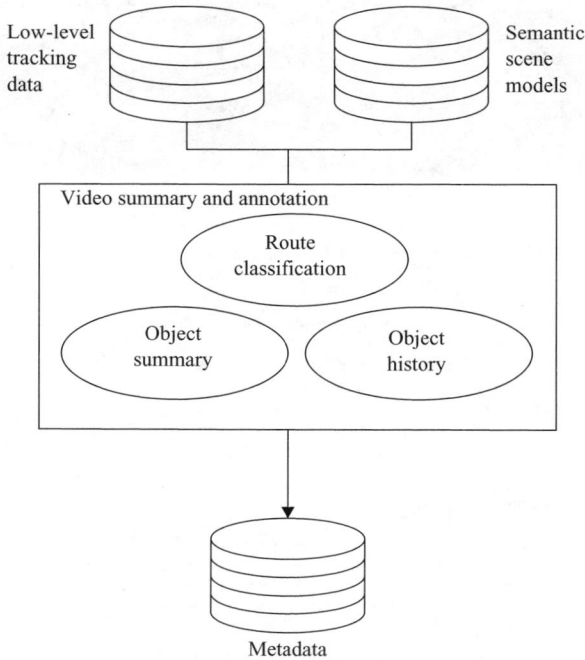

Figure 3.10   The data flow for online metadata generation

appropriately to replay the tracking video. The current system supports the following types of activity queries:

- Select all objects that have used a specific entry or exit zone.
- Retrieve all objects that have used a specific route in a specific camera.
- Retrieve all objects that have visited a combination of routes in a specific order.
- Retrieve all objects that have passed through a section of a route.
- Count the route popularity by time of day.

Figure 3.11 illustrates how the database is used to perform route classification for two of the tracked object trajectories. Four routes are shown that are stored in the semantic description layer of the database in Figure 3.11(a). In this instance the

(a)

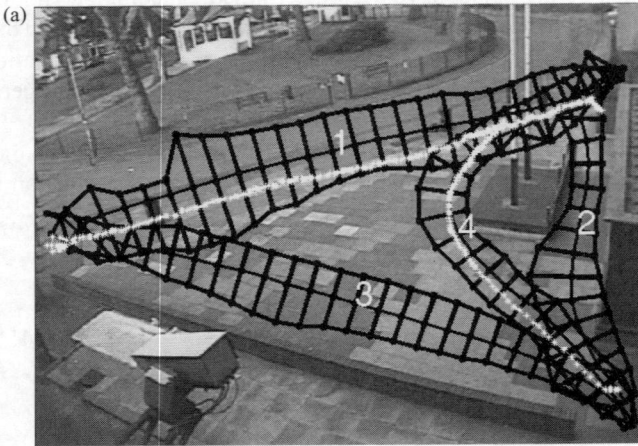

(b)

```
select routeid, count(nodeid)
from routenodes r, objects o
where camera=2
and o.path ?# r.polyzone
and o.videoseq=87
and o.trackid in (1,3)
group by routeid
```

(c)

Videoseq	Trackid	Start Time	End Time	Route
87	1	08:16:16	08:16:27	4
87	3	08:16:31	08:16:53	1

Figure 3.11   (a) Example of route classification; (b) SQL query to find routes that intersect with an object trajectory; (c) the object history information for both of the tracked objects

```
Select o.videoseq, o.trackid, o.start_time
from object_summary o, entryzone e, exitzone x
where o.start_pos && e.region
and o.end_pos && x.region
and e.label = 'B' and x.label in ('A','C')
and o.start_time between '2004-05-24 10:30' and '2004-05-24 11:00'
```

Figure 3.12   SQL query to retrieve all objects moving between entry zone B and exit zones (A,C)

first object trajectory is assigned to route 4, since this is the route with the largest number of intersecting nodes. The second object trajectory is assigned to route 1. The corresponding SQL query used to classify routes is shown in Figure 3.11(b). Each node along the route is modelled as a polygon primitive provided by the PostgreSQL database engine. The query counts the number of route nodes that intersects with the object's trajectory. This allows a level of discrimination between ambiguous choices for route classification. The '?#' operator in the SQL statement is a logical operator that returns true if the object trajectory intersects with the polygon region of a route node. Additional processing of the query results allows the system to generate the history of the tracked object in terms of the route models stored in the semantic description layer. A summary of this information generated for the two displayed trajectories is given in Figure 3.11(c). It should be noted that if a tracked object traversed multiple routes during its lifetime then several entries would be created for each route visited.

In Figure 3.12 the activity query used to identify objects moving between a specific entry and exit zone is shown. The query would return all objects entering the FOV by entry zone B and leaving the FOV by either exit zone A or exit zone C. The underlying database table 'object_summary' is metadata that provides an annotated summary of all the objects detected within the FOV. The '&&' function in the SQL statement is a geometric operator in PostgreSQL that determines if the object's entry or exit position is within a given entry or exit region. The last clause at the end of the activity query results in only objects detected between 10:30 a.m. and 11:00 a.m. on May 24, 2003 being returned in the output. This example demonstrates how the metadata can be used to generate spatial–temporal queries of object activity. Figure 3.13 shows the tracked objects returned by the activity query in Figure 3.12.

The metadata provides better indexing of the object motion and image framelet layers of the database, which results in improved performance for various types of activity queries. The use of spatial–temporal SQL would allow a security operator to construct the following types of queries:

- Retrieve all objects moving within a restricted area over a specific time period.
- Retrieve a list of all objects leaving a building within a specific time period.
- Retrieve a list of all objects entering a building within a specific time period.
- Show examples of two or more objects moving along a route simultaneously.

Figure 3.13   *Sample of results returned by spatial–temporal activity queries: objects moving from entry zone B to exit zone A, and objects moving from entry zone B to exit zone C*

The response times of spatial–temporal queries can be reduced from several minutes or hours to only a few seconds. This is due to the metadata being much lighter to query when compared to the original video. This is of practical use for real-world applications where a security operator may be required to review and browse the content of large volumes of archived CCTV footage.

3.6   Performance evaluation

A typical approach to evaluating the performance of the detection and tracking system uses ground truth to provide independent and objective data (e.g. classification, location, size) that can be related to the observations extracted from the video sequence. Manual ground truth is conventionally gathered by a human operator who uses a 'point and click' user interface to step through a video sequence and select well-defined points for each moving object. The manual ground truth consists of a set of points that define the trajectory of each object in the video sequence (e.g. the object centroid). The human operator decides if objects should be tracked as individuals or classified as a group. The motion detection and tracking algorithm is then run on the pre-recorded video sequence and ground truth and tracking results are compared to assess tracking performance.

The reliability of the video tracking algorithm can be associated with a number of criteria: the frequency and complexity of dynamic occlusions, the duration of

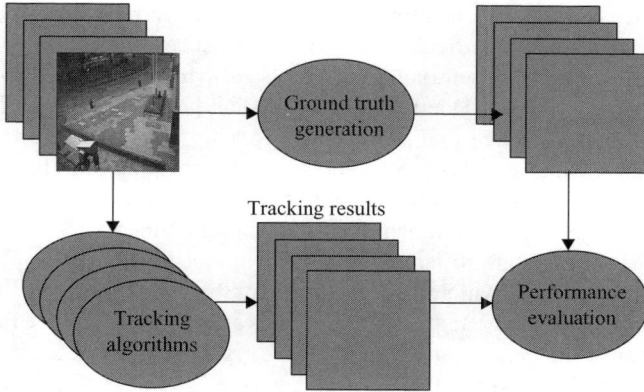

Figure 3.14   Generic framework for quantitative evaluation of a set of video tracking algorithm.

targets behind static occlusions, the distinctiveness of the targets (e.g. if they are all different colours) and changes in illumination or weather conditions. In this chapter we express a measure for estimating the perceptual complexity of the sequence based on the occurrence and duration of dynamic occlusions, since this is the event most likely to cause the tracking algorithm to fail. Such information can be estimated from the ground truth data by computing the ratio of the number of target occlusion frames divided by the total length of each target track (i.e. the number of frames over which it is observed), averaged over the sequence.

A general framework for quantitative evaluation of a set of video tracking algorithms is shown in Figure 3.14. Initially a set of video sequences must be captured in order to evaluate the tracking algorithms. Ideally, the video sequences should represent a diverse range of object tracking scenarios, which vary in perceptual complexity to provide an adequate test for the tracking algorithms. Once the video data have been acquired, ground truth must then be generated to define the expected tracking results for each video sequence. Ground truth, as previously discussed, can consist of the derived trajectory of the centroid of each object along with other information such as the bounding box dimensions. Given the complete set of ground truth for the testing datasets, the tracking algorithms are then applied to each video sequence, resulting in a set of tracking results. The tracking results and the ground truth are then compared in order to measure the tracking performance using an appropriate set of surveillance metrics.

One of the main issues with the quantitative evaluation of video tracking algorithms is that it can be time consuming to acquire an appropriate set of video sequences that can be used for performance evaluation. This problem is being partially addressed by the Police Scientific Development Branch (PSDB), who are gathering data for a Video Test Image Library (VITAL) [24], which will represent a broad range of object tracking and surveillance scenarios encompassing parked vehicle detection, intruder detection, abandoned baggage detection, doorway surveillance and abandoned vehicles. Generating ground truth for pre-recorded video can be a time-consuming process, particularly for video sequences that contain a large number of objects. A number of semi-automatic tools are available to speed up the process of ground truth generation. Doermann and Mihalcik [25] created the Video Performance Evaluation Resource (ViPER) to provide a software interface that could be used to visualise video analysis results and metrics for evaluation. The interface was developed in Java and is publicly available for download. Jaynes *et al.* developed the Open Development Environment for Evaluation of Video Surveillance Systems (ODViS) [26]. The system differs from ViPER in that it offers an application programmer interface (API), which supports the integration of new surveillance modules into the system. Once integrated, ODViS provides a number of software functions and tools to visualise the behaviour of the tracking systems. The integration of several surveillance modules into the ODViS framework allows several different tracking algorithms to be compared to each other, or pre-defined ground truth. The development of the ODViS framework is an ongoing research effort. Plans are underway to support a variety of video formats and different types of tracking algorithms. The ViPER and ODViS frameworks provide a set of software tools to capture ground truth and visualise tracking results from pre-recorded video. In Reference 27 ground truth is automatically generated by using pre-determined cues such as shape and size on controlled test sequences. Once the ground truth is available there are a number of metrics that can be applied in order to measure tracking performance [22,25,28–31]. At this point in time there is no automatic method available for measuring the perceptual complexity of each video sequence, and automatically capturing ground truth for each dataset. An ideal solution for the generic quantitative evaluation framework depicted in Figure 3.14 would fully automate the video data and ground truth acquisition steps. If these steps were fully automated it would be practical to evaluate tracking algorithms over very large volumes of test data, which would not be feasible with either manual or semi-automatic methods.

3.7   Pseudo-synthetic video

As an alternative to manual ground truthing, we propose using pseudo-synthetic video to evaluate tracking performance. A problem for performance evaluation of tracking algorithms is that it is not trivial to accumulate datasets of varying perceptual complexity. Ideally, we want to be able to run a number of experiments and vary the perceptual complexity of the scene to test the tracking algorithm under a variety of different conditions. This is possible using manual ground truth but requires the capture of a large number of video sequences, which may not be practical at some surveillance sites.

The novelty of our framework is that we automatically compile a set of isolated ground truth tracks from the surveillance database. We then use the ground truth tracks to construct a comprehensive set of pseudo-synthetic video sequences that are used to evaluate the performance of a tracking algorithm.

3.7.1   *Ground truth track selection*

A list of ground truth tracks is initially compiled from the surveillance database. We select ground truth tracks during periods of low object activity (e.g. over weekends), since there is a smaller likelihood of object interactions that can result in tracking errors. The ground truth tracks are checked for consistency with respect to path coherence, colour coherence, and shape coherence in order to identify and remove tracks of poor quality.

Path coherence

The path coherence metric [22] makes the assumption that the derived tracked object trajectory should be smooth subject to direction and motion constraints. Measurements are penalised for lower consistency with respect to direction and speed, while measurements are rewarded for the converse situation.

$$
\varepsilon_{pc} = \frac{1}{N-2} \sum_{k=2}^{N-1} \left\{ w_1 \left(1 - \frac{\overline{X_{k-1}X_k} \cdot \overline{X_kX_{k+1}}}{\|\overline{X_{k-1}X_k}\| \|\overline{X_kX_{k+1}}\|} \right) \right.
$$
$$
\left. + w_2 \left(1 - \frac{2\sqrt{\|\overline{X_{k-1}X_k}\| \|\overline{X_kX_{k+1}}\|}}{\|\overline{X_{k-1}X_k}\| + \|\overline{X_kX_{k+1}}\|} \right) \right\},
$$

where $\overline{X_{k-1}X_k}$ is the vector representing the positional shift of the tracked object between frames k and $k-1$. The weighting factors can be appropriately assigned to define the contribution of the direction and speed components of the measure. The value of both weights was set to 0.5.

Colour coherence

The colour coherence metric measures the average inter-frame histogram distance of a tracked object. It is assumed that the object histogram should remain constant between image frames. The normalised histogram is generated using the normalised (r,g) colour space in order to account for small lighting variations. This metric has low values if the segmented object has similar colour attributes and higher values when colour attributes are different. Each histogram contains 8×8 bins for the normalised colour components.

$$
\varepsilon_{cc} = \frac{1}{N-1} \sum_{k=2}^{N} \sqrt{1 - \sum_{u=1}^{M} p_{k-1}(u) p_k(u)},
$$

where $p_k(u)$ is the normalised colour histogram of the tracked object at frame k, which has M bins and N is the number of frames the object was tracked over. This metric is a popular colour similarity measure employed by several robust tracking algorithms [32,33].

Shape coherence

The shape coherence metric gives an indication of the level of agreement between the tracked object position and the object foreground region. This metric will have a high value when the localisation of the tracked object is incorrect due to poor initialisation or an error in tracking. The value of the metric is computed by evaluating the symmetric shape difference between the bounding box of the foreground object and tracked object state.

$$\varepsilon_{sc} = \frac{1}{N} \sum_{k=1}^{N} \frac{|R_f(k) - R_t(k)| + |R_t(k) - R_f(k)|}{|R_t(k) \cup R_f(k)|},$$

where $|R_t(k) - R_f(k)|$ represents the area difference between the bounding box of the tracked object (state) and the overlapping region with the foreground object (measurement). The normalisation factor $|R_t(k) \cup R_f(k)|$ represents the area of the union of both bounding boxes.

Outlier ground truth tracks can be removed by applying a threshold to the values of ε_{pc}, ε_{cc} and ε_{sc}. The distributions of the values are shown in Figure 3.15. It can be observed that each metric follows a Gaussian distribution. The threshold is set so that the value should be within two standard deviations of the mean. The mean and standard deviations of ε_{pc}, ε_{cc} and ε_{sc} were (0.092, 0.086, 0.157) and (0.034, 0.020, 0.054), respectively. We also exclude any tracks that are short in duration and have not been tracked for at least $N = 50$ frames or have formed a dynamic occlusion with another track.

In Figure 3.16 some example outlier tracks are shown. The top left track was rejected due to poor path coherence, since the derived object trajectory is not smooth. The top right track was rejected due to low colour coherence, which is a consequence of the poor object segmentation. The bottom left track was rejected due to poor shape coherence, where an extra pedestrian merges into the track. The tracked bounding boxes are not consistent with the detected foreground object. The bottom right track was rejected due to forming a dynamic occlusion with another track. It can be observed that in this instance the tracking failed and the objects switched identities near the bottom of the image. These examples illustrate that the metrics: path coherence, colour coherence, and shape coherence are effective for rejecting outlier ground truth tracks of poor quality.

3.7.2   *Pseudo-synthetic video generation*

Once the ground truth tracks have been selected they are employed to generate pseudo-synthetic videos. Each pseudo-synthetic video is constructed by replaying the ground truth tracks randomly in the generated video sequence. Two ground truth tracks are shown in left and middle images of Figure 3.17, the tracked object is plotted every few frames in order to visualise the motion history of the object through the scene. When the two tracks are inserted in a pseudo-synthetic video sequence a dynamic occlusion can be created as shown in the right image of Figure 3.17. Since the ground truth is known for each track we can determine the exact time and duration of the dynamic occlusion. By adding more ground truth tracks more complex object interactions are generated.

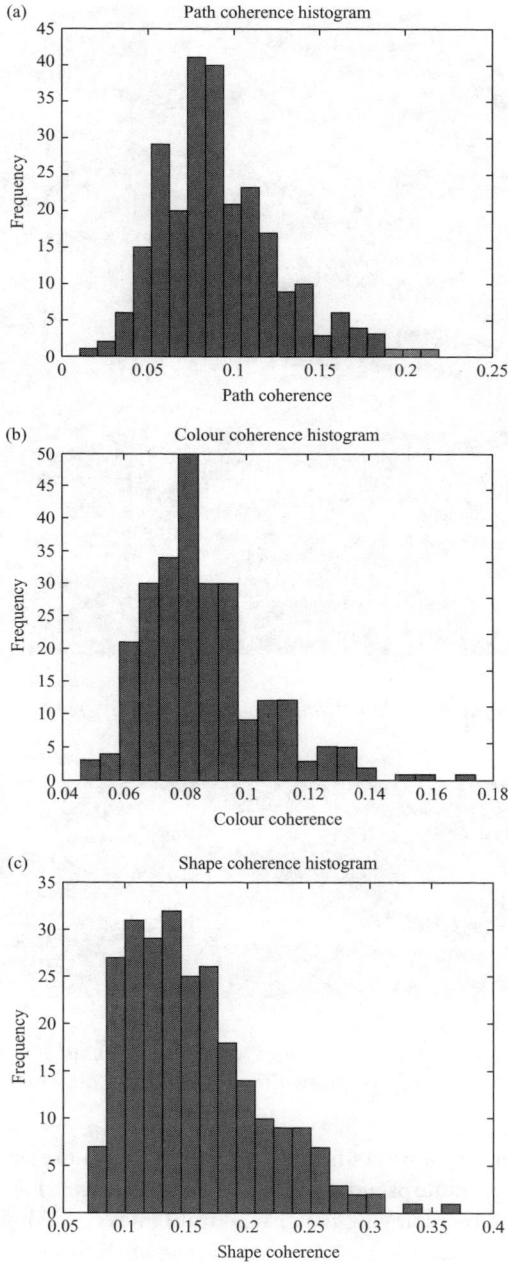

Figure 3.15   Distribution of the average path coherence (a), average colour coherence (b) and average shape coherence of each track selected from the surveillance database

Figure 3.16   Example of outlier tracks identified during ground truth track selection

Figure 3.17   The left and middle images show two ground truth tracks. The right image shows how the two tracks can form a dynamic occlusion

Once the ground truth tracks have been selected from the surveillance database they can be used to generate pseudo-synthetic video sequences. Each pseudo-synthetic video is constructed by replaying the ground truth tracks randomly in the generated video sequence. One issue associated with this approach is that few ground truth tracks will be observed in regions where tracking or detection performance is poor. However, the pseudo-synthetic data are still effective for characterising tracking performance with respect to tracking correctness and dynamic occlusion reasoning, which is the main focus of the evaluation framework. A fundamental point of the method is that the framelets stored in the surveillance database consist of the foreground regions identified by the motion detection algorithm (i.e. within the bounding box). When the framelet is re-played in the pseudo-synthetic video sequence this improves the realism of the dynamic occlusions. A number of steps are taken to construct each pseudo-synthetic video sequence, since the simple insertion of ground truth tracks would not be sufficient to create a realistic video sequence.

- Initially, a dynamic background video is captured for the same camera view from which the ground truth tracks have been selected. This allows the pseudo-synthetic video to simulate small changes in illumination that typically occur in outdoor environments.
- All the ground truth tracks are selected from the same fixed camera view. This ensures that the object motion in the synthetic video sequence is consistent with background video. In addition, since the ground truth tracks are constrained to move along the same paths in the camera view, this increases the likelihood of forming dynamic occlusions in the video sequences.
- 3D track data are used to ensure that framelets are plotted in the correct order during dynamic occlusions, according to their estimated depth from the camera. This gives the effect of an object occluding or being occluded by other objects based on its distance from the camera.

There are several benefits of using pseudo-synthetic video: it is possible to simulate a wide variety of dynamic occlusions of varying complexity; pseudo-synthetic video can be generated for a variety of weather conditions; the perceptual complexity of each synthetic video can be automatically estimated; and ground truth can be automatically acquired. One disadvantage is that the pseudo-synthetic video is biased towards the motion detection algorithm used to capture the original data, and few ground truth tracks will be generated in regions where tracking or detection performance is poor. In addition, the metrics described in Section 3.8.1 do not completely address all the problems associated with motion segmentation, for example, the effects of shadows cast by moving objects, changes in weather conditions, the detection of low contrast objects and the correct segmentation of an object's boundary. However, the pseudo-synthetic video is effective for evaluating the performance of tracking with respect to dynamic occlusion reasoning, which is the main focus of the performance evaluation framework.

3.7.3 Perceptual complexity

The perceptual complexity of each pseudo-synthetic video sequence is controlled by a set of tuneable parameters:

Max Objects (Max). The maximum number of the objects that can be present in any frame of the generated video sequence.

New Object Probability – p(new). The probability of creating a new object in the video sequence while the maximum number of objects has not been exceeded.

Figure 3.18   Perceptual complexity: left – p(new) = 0.01 image; middle – framelets plotted for p(new) = 0.1; right – framelets plotted for p(new) = 0.2

Increasing the value of p(new) results in a larger number of objects appearing in the constructed video sequence. This is illustrated in Figure 3.18 where the three images demonstrate how the value of p(new) can be used to control the density of objects in each pseudo-synthetic video sequence. The images show examples for p(new) having the values 0.01, 0.10 and 0.20, respectively. These two parameters are used to vary the complexity of each generated video sequence. Increasing the values of the parameters results in an increase in object activity. We have found this model provides a realistic simulation of actual video sequences.

The number of dynamic occlusions in each pseudo-synthetic video was determined by counting the number of occurrences where the bounding box of two or more ground truth objects overlaps in the same image frame. We can count the number of dynamic occlusions (NDO), the average number of occluding objects (NOO), and the average duration of a dynamic occlusion (DDO) to provide a measure of the perceptual complexity [22]. Figure 3.19 demonstrates how p(new) can vary the perceptual complexity of each generated pseudo-synthetic video. Figure 3.19(a) and (b) are plots of p(new) by average number of objects per frame in the pseudo-synthetic video, and the average number of dynamic object occlusions, respectively. The error bars on each plot indicate the standard deviation over the five simulations performed for each value of p(new). The values become asymptotic in both plots as the number of objects per frame approaches the maximum of 20, representing a complex and dense video sequence.

3.7.4   Surveillance metrics

Once the tracking algorithm has been used to process each generated pseudo-synthetic video sequence, the ground truth and tracking results are compared to generate a surveillance metrics report. The surveillance metrics report has been derived from a number of sources [25,28–31]. Minimising the following trajectory distance measure allows us to align the objects in the ground truth and tracking results:

$$D_T(g,r) = \frac{1}{N_{rg}} \sum_{\exists ig(t_i) \wedge r(t_i)} \sqrt{(xg_i - xr_i)^2 + (yg_i - yr_i)^2},$$

where N_{rg} is the number of frames that the ground truth track and result track have in common, and (xg_i, yg_i), (xr_i, yr_i) are the location of the ground truth and result

(a)

(b)

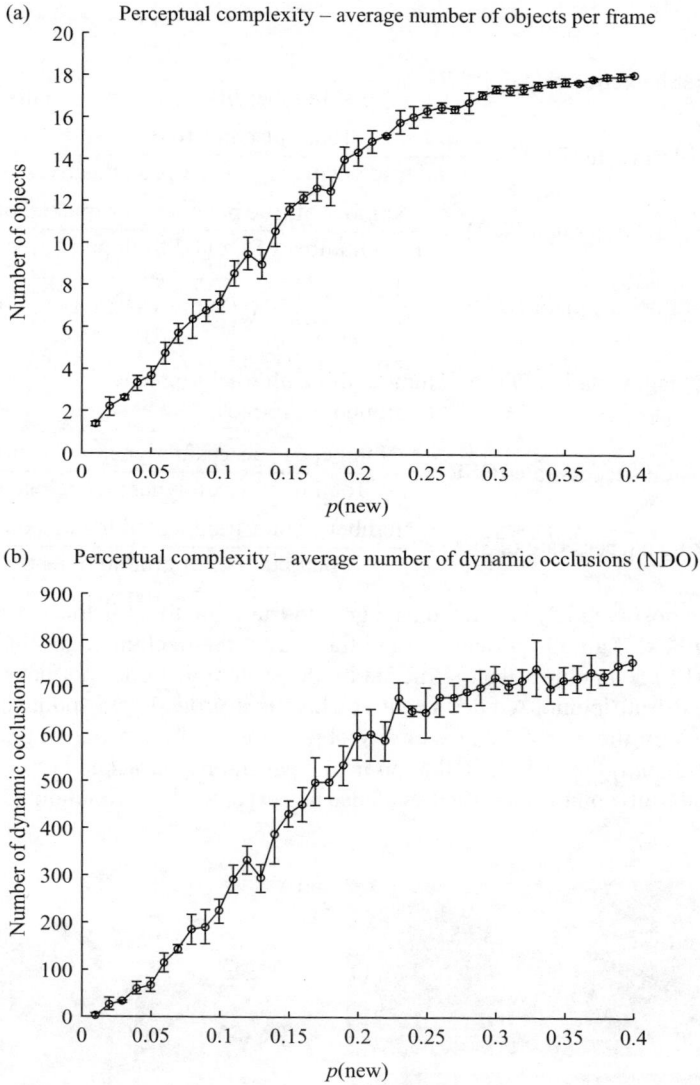

Figure 3.19   *(a) Plot of average no. of objects per frame by p(new); (b) plot of average no. of dynamic occlusions by p(new). No. of frames = 1500, max no. of objects = 20*

object at frame i, respectively. Once the ground truth and results trajectories have been matched, the following metrics are used in order to characterise the tracking performance:

$$\text{Tracker detection rate (TRDR)} = \frac{\text{Total true positives}}{\text{Total number of ground truth points}},$$

$$\text{False alarm rate (FAR)} = \frac{\text{Total false positives}}{\text{Total true positives} + \text{Total false positives}},$$

$$\text{Track detection rate (TDR)} = \frac{\text{Number of true positives for tracked object}}{\text{Total number of ground truth points for object}},$$

$$\text{Object tracking error (OTE)} = \frac{1}{N_{rg}} \sum_{\exists ig(t_i) \wedge r(t_i)} \sqrt{(xg_i - xr_i)^2 + (yg_i - yr_i)^2},$$

$$\text{Track fragmentation (TF)} = \text{Number of result tracks matched to ground truth track},$$

$$\text{Occlusion success rate (OSR)} = \frac{\text{Number of successful dynamic occlusions}}{\text{Total number of dynamic occlusions}},$$

$$\text{Tracking success rate (TSR)} = \frac{\text{Number of non-fragmented tracked objects}}{\text{Total number of ground objects}}.$$

A true positive (TP) is defined as a ground truth point that is located within the bounding box of an object detected and tracked by the tracking algorithm. A false positive (FP) is an object that is tracked by the system that does not have a matching ground truth point. A false negative (FN) is a ground truth point that is not located within the bounding box of any object tracked by the tracking algorithm. In Figure 3.20(a) the vehicle in the top image has not been tracked correctly, hence the ground truth point is classified as a false negative, while the bounding box of the

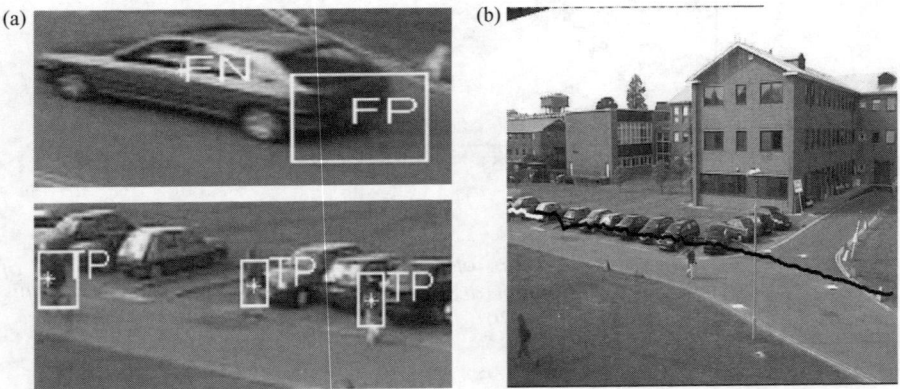

Figure 3.20   Illustration of surveillance metrics: (a) image to illustrate true positives, false negative and false positive; (b) image to illustrate a fragmented tracked object trajectory

incorrectly tracked object is counted as a false positive. The three objects in the bottom image are counted as true positives, since the ground truth points are located within the tracked bounding boxes.

The tracker detection rate (TRDR) and false alarm rate (FAR) characterise the performance of the object-tracking algorithm. The track detection rate (TDR) indicates the completeness of a specific ground truth object. The object tracking error (OTE) indicates the mean distance between the ground truth and tracked object trajectories. The track fragmentation (TF) indicates how often a tracked object label changes. Ideally, the TF value should be 1, with larger values reflecting poor tracking and trajectory maintenance. The tracking success rate (TSR) summarises the performance of the tracking algorithm with respect to track fragmentation. The occlusion success rate (OSR) summarises the performance of the tracking algorithm with respect to dynamic occlusion reasoning.

Figure 3.20(b) shows a tracked object trajectory for the pedestrian who is about to leave the camera field of view. The track is fragmented into two parts shown as black and white trajectories. The two track segments are used to determine the track detection rate, which indicates the completeness of the tracked object. As a result this particular ground truth object had a TDR, OTE and TF of 0.99, 6.43 pixels and 2, respectively.

3.8 Experiments and evaluation

3.8.1 Single-view tracking evaluation (qualitative)

A number of experiments were run to test the performance of the tracking algorithm used by the online surveillance system. The tracking algorithm employs a partial observation tracking model [23] for occlusion reasoning. Manual ground truth was generated for the PETS2001 datasets using the point and click graphical user interface as described in Section 3.6. The PETS2001 dataset is a set of video sequences that have been made publicly available for performance evaluation. Each dataset contains static camera views with pedestrians, vehicles, cyclists and groups of pedestrians. The datasets were processed at a rate of 5 fps, since this approximately reflects the operating speed of our online surveillance system. Table 3.1 provides a summary of the surveillance metrics reports. The results demonstrate the robust tracking performance, since the track completeness is nearly perfect for all the objects. A couple of the tracks are fragmented due to poor initialisation or termination. Figure 3.21 demonstrates what can happen when a tracked object is not initialised correctly. The left, middle and right images show the pedestrian exiting and leaving the parked vehicle. The pedestrian is partially occluded by other objects and so is not detected by the tracking algorithm until he has moved from the vehicle. The pedestrian relates to ground truth object 9 in Table 3.1.

An example of dynamic occlusion reasoning is shown in Figure 3.22. The cyclist overtakes the two pedestrians, forming two dynamic occlusions, and it can be noted that the correct trajectory is maintained for all three objects. The object labels in Figure 3.22 have been assigned by the tracking algorithm and are different from the

Table 3.1   Summary of surveillance metrics for PETS2001 dataset 2, camera 2

Track	0	1	2	3	4	5	6	7	8	9
TP	25	116	26	104	36	369	78	133	43	88
FN	0	2	0	5	0	5	1	1	1	2
TDR	1.00	0.98	1.00	0.95	1.00	0.99	0.99	0.99	0.98	0.98
TF	1	1	1	1	1	1	1	2	1	2
OTE	11.09	7.23	8.37	4.70	10.82	11.63	9.05	6.43	8.11	11.87

TP: number of true positives.
FN: number of false positives.
TF: track fragmentation.
TDR: track detection rate.
OTE: object tracking error.

Figure 3.21   An example of how poor track initialisation results in a low object track detection rate of the pedestrian leaving the vehicle

ground truth object labels. Table 3.2 gives a summary of the tracking performance of the PETS2001 datasets. These results validate our assumption that our object tracker can be used to generate ground truth for video with low activity.

The PETS datasets were used to construct a pseudo-synthetic video by adding four additional ground truth tracks to the original sequence. Table 3.3 summarises the differences in perceptual complexity between the original and synthetic video sequence. The number of dynamic object occlusions increases from 4 to 12, having the desired effect of increasing the complexity of the original video sequence.

3.8.2   Single-view tracking evaluation (quantitative)

To test the effectiveness of the tracking algorithm for tracking success and dynamic occlusion reasoning, a number of synthetic video sequences were generated. Initially, ground truth tracks were automatically selected from a surveillance database using the method described in Section 3.7.3. The tracks in the surveillance database were observed over a period of 8 h by a camera overlooking a building entrance. Five pseudo-synthetic video sequences were then generated for each level of perceptual complexity. The value of $p(\text{new})$ was varied between 0.01 and 0.4 with increments of 0.01. Each synthetic video sequence was 1500 frames in length, which is equivalent to approximately 4 min of live captured video by our online system running at 7 Hz.

Figure 3.22   Example of dynamic occlusion reasoning for PETS2001 dataset 2, camera 2

Table 3.2   Summary of object tracking metrics

	TRDR	TSR	FAR	AOTE		ATDR	
				Mean	Stdev	Mean	Stdev
Dataset 2 (Cam 2)	0.99	8/10	0.01	8.93	2.40	0.99	0.010
Pseudo-synthetic PETS dataset	1.00	9/13	0.01	1.36	2.09	1.00	0.002

Hence, in total the system was evaluated with 200 different video sequences, totalling approximately 800 min of video with varying perceptual complexity. The synthetic video sequences were used as input to the tracking algorithm.

The tracking results and ground truth were then compared and used to generate a surveillance metrics report as described in Section 3.7.4. Table 3.4 gives a summary of the complexity of a selection of the synthetic video sequences. These results confirm that p(new) controls the perceptual complexity, since the number of objects, average number of dynamic occlusions and occluding objects increases from (19.4, 6.2, 2.1) to (326.6, 761.4, 3.1), respectively, for the smallest and largest values of p(new).

Table 3.3   Summary of perceptual complexity of PETS datasets

	TNO	NDO	DDO	NOO
Original PETS dataset	10	4	8.5	1
Pseudo-synthetic PETS dataset	14	12	8.58	1.08

NDO: number of dynamic occlusions.
DDO: duration of dynamic occlusion (frames).
NOO: number of occluding objects.
TNO: total number of objects.

Table 3.4   Summary of the perceptual complexity of the synthetic video sequences

p(new)	TNO		NDO		DDO		NOO	
	Mean	Stdev	Mean	Stdev	Mean	Stdev	Mean	Stdev
0.01	19.40	1.673	6.20	1.304	10.59	4.373	2.08	0.114
0.20	276.40	7.668	595.20	41.889	11.51	1.314	2.95	0.115
0.40	326.60	13.240	761.40	49.958	11.53	0.847	3.06	0.169

Table 3.5   Summary of metrics generated using each synthetic video sequence

p(new)	TRDR	FAR	OSR		AOTE		ATDR		ATSR	
			Mean	Stdev	Mean	Stdev	Mean	Stdev	Mean	Stdev
0.01	0.91	0.052	0.77	0.132	3.43	0.582	0.89	0.070	0.81	0.080
0.20	0.86	0.009	0.56	0.021	12.49	0.552	0.74	0.011	0.21	0.023
0.40	0.85	0.006	0.57	0.021	13.26	0.508	0.73	0.007	0.19	0.020

Table 3.5 summarises the tracking performance for various values of p(new). The plot in Figure 3.23 demonstrates how the object tracking error (OTE) varies with the value of p(new). Large values of p(new) result in an increase in the density and grouping of objects; as a result, this causes the tracking error of each object to increase.

The plot in Figure 3.24 illustrates how the tracker detection rate (TDR) varies with the value of p(new). The value of the TDR decreases from 90 to 72 per cent with increasing p(new). This indicates that the tracking algorithm has a high detection rate for each object with increasing perceptual complexity of the video sequence. This result is expected since we expect that the video sequences generated would

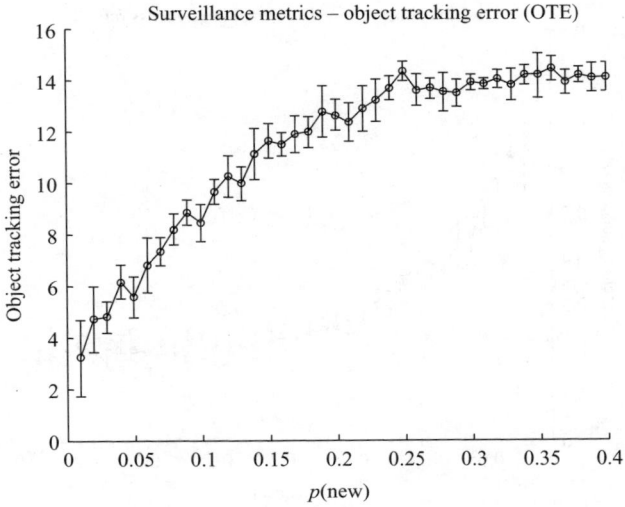

Figure 3.23   Plot of object tracking error (OTE)

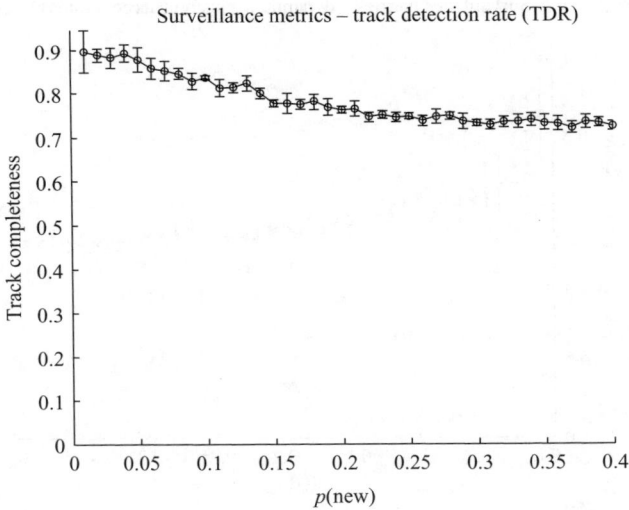

Figure 3.24   Plot of track detection rate (TDR)

have a degree of bias towards the motion segmentation algorithm used to capture and store the ground truth tracks in the surveillance database. The TDR indicates how well an object has been detected but does not account for track fragmentation and identity switching, which can occur during a dynamic occlusion that is not correctly resolved by the tracker. This is more accurately reflected by the occlusion

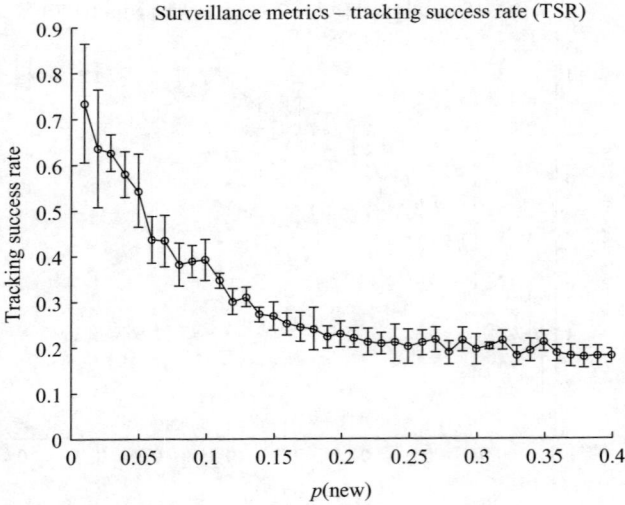

Figure 3.25   Plot of tracking success rate (TSR)

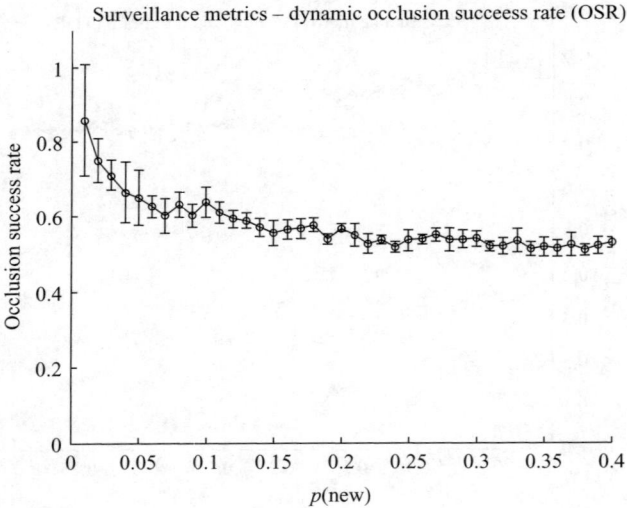

Figure 3.26   Plot of occlusion success rate (OSR)

success rate (OSR) and the tracking success rate (TSR) metrics, which are shown in Figures 3.25 and 3.26. The track fragmentation increases with the value of p(new), which represents a degradation of tracking performance with respect to occlusion reasoning. The OSR and TSR decreases in value from (77 and 81 per cent) to (57 and 19 per cent) with increasing value of p(new). When the number of objects per frame approaches the maximum, this limits the number of dynamic occlusions created; hence increasing values of p(new) have a diminished effect of increasing the perceptual complexity. As a consequence the OTE, TDR, TSR and OSR become asymptotic once the number of objects per frame approaches the maximum of 20 as illustrated in the plots of Figures 3.25 and 3.26. Larger values of p(new) and the maximum number of objects should result in more complex video sequences. Hence, even with the bias present in the generated video sequences we can still evaluate the object tracking performance with respect to tracking success and occlusion reasoning, without exhaustive manual ground truthing, fulfilling the main objective of our framework for performance evaluation.

3.9 Summary

We have presented a distributed surveillance system that can be used for continuous monitoring over extended periods, and a generic quantitative performance evaluation framework for video tracking. We have demonstrated how a surveillance database can support both the operational and reporting requirements of a CCTV system. In future work we plan to make the metadata MPEG-7 compliant [34]. MPEG-7 is a standard for video content description that is expressed in XML. MPEG-7 only describes content and is not concerned with the methods used to extract features and property from the originating video. By adopting these standards we can future proof our surveillance database and ensure compatibility with other content providers.

References

1 Q. Cai and J.K. Aggarwal. Tracking human motion in structured environments using a distributed camera system. *IEEE Pattern Analysis and Machine Intelligence (PAMI)*, 1999;2(11):1241–1247.
2 T. Chang and S. Gong. Tracking multiple people with a multi-camera system. In *IEEE Workshop on Multi-Object Tracking (WOMOT01)*, Vancouver, Canada, July 2001, pp. 19–28.
3 R.T. Collins, A.J. Lipton, H. Fujiyoshi, and T. Kanade. Algorithms for cooperative multisensor surveillance. *Proceedings of the IEEE*, 2001;89(10):1456–1476.
4 O. Javed and M. Shah. KNIGHT: a multi-camera surveillance system. In *IEEE International Conference on Multimedia and Expo 2003*, Baltimore, July 2003.
5 G.T. Kogut and M.M. Trivedi. Maintaining the identity of multiple vehicles as they travel through a video network. In *IEEE Workshop on Multi-Object Tracking (WOMOT01)*, Vancouver, Canada, July 2001, pp. 29–34.
6 L. Lee, R. Romano, and G. Stein. Monitoring activities from multiple video streams: establishing a common coordinate frame. *IEEE Pattern Analysis and Machine Intelligence (PAMI)*, 2000;22(8):758–767.
7 P. Remagnino and G.A. Jones. Classifying surveillance events and attributes and behavior. In *British Machine Vision Conference (BMVC2001)*, Manchester, September 2001, pp. 685–694.

8 G. Medioni, I. Cohen, F. Bremod, S. Hongeng, and R. Nevatia. Event detection and analysis from video streams. *IEEE Pattern Analysis and Machine Intelligence (PAMI)*, 2001;23(8):873–889.

9 S. Guler, W.H. Liang, and I. Pushee. A video event detection and mining framework. In *IEEE Workshop on Event Mining Detection and Recognition of Events in Video*, June 2003.

10 Y.A. Ivanov and A.F. Bobick. Recognition of visual activities and interactions by stochastic parsing. *IEEE Pattern Analysis and Machine Intelligence (PAMI)*, 2000;22(8):852–872.

11 B. Katz, J. Lin, C. Stauffer, and E. Grimson. Answering questions about moving objects in surveillance videos. In *Proceedings of the 2003 Spring Symposium on New Directions in Question Answering*, March 2003.

12 M. Trivedi, S. Bhonsel, and A. Gupta. Database architecture for autonomous transportation agents for on-scene networked incident management (ATON). In *International Conference on Pattern Recognition (ICPR)*, Barcelona, Spain, 2000, pp. 4664–4667.

13 X. Cai, F. Ali, and E. Stipidis. *MPEG-4 Over Local Area Mobile Surveillance System*. In IEE Intelligent Distributed Surveillance Systems, London, UK, February 2003.

14 D. Makris and T. Ellis. Finding paths in video sequences. In *British Machine Vision Conference BMVC2001*, University of Manchester, UK, Vol. 1, September 2001, pp. 263–272.

15 D. Makris and T. Ellis. Automatic learning of an activity-based semantic scene model. In *IEEE Conference on Advanced Video and Signal Based Surveillance (AVSB 2003)*, Miami, July 2003, pp. 183–188.

16 D. Makris and T. Ellis. Path detection in video surveillance. *Image and Vision Computing Journal*, 2000;20:895–903.

17 D. Makris and T. Ellis. Spatial and probabilistic modelling of pedestrian behaviour. In *British Machine Vision Conference (BMVC2002)*, Cardiff University, UK, Vol. 2, September 2002, pp. 557–566.

18 D. Makris, T.J. Ellis, and J. Black. Bridging the gaps between cameras. In *IEEE Computer Society Conference on Computer Vision and Pattern Recognition CVPR 2004*, Washington DC, June 2004, Vol. 2, pp. 205–210

19 J. Black and T.J. Ellis. Multi-view image surveillance and tracking. In *IEEE Workshop on Motion and Video Computing*, Orlando, December 2002, pp. 169–174.

20 J. Black, T.J. Ellis, and D. Makris. A hierarchical database for visual surveillance applications. In *IEEE International Conference on Multimedia and Expo (ICME2004)*, June, Taipei, Taiwan, 2004.

21 M. Xu and T.J. Ellis. Illumination-invariant motion detection using color mixture models. In *British Machine Vision Conference (BMVC 2001)*, Manchester, September 2001, pp. 163–172.

22 J. Black, T.J. Ellis, and P. Rosin. A novel method for video tracking performance evaluation. In *The Joint IEEE International Workshop on Visual Surveillance and Performance Evaluation of Tracking and Surveillance (VS-PETS)*, Nice, France, October 2003, pp. 125–132.

23 M. Xu and T.J. Ellis. Partial observation vs blind tracking through occlusion. In *British Machine Vision Conference (BMVC 2002)*, Cardiff, September 2002, pp. 777–786.

24 D. Baker. *Specification of a Video Test Imagery Library (VITAL)*. In IEE Intelligent Distributed Surveillance Systems, London, February 2003.

25 D. Doermann and D. Mihalcik. Tools and techniques for video performance evaluation. In *Proceedings of the International Conference on Pattern Recognition (ICPR'00)*, Barcelona, September 2000, pp. 4167–4170.

26 C. Jaynes, S. Webb, R. Matt Steele, and Q. Xiong. An open development environment for evaluation of video surveillance systems. In *Proceedings of the Third International Workshop on Performance Evaluation of Tracking and Surveillance (PETS'2002)*, Copenhagen, June 2002, pp. 32–39.

27 P.L. Rosin and E. Ioannidis. Evaluation of global image thresholding for change detection. *Pattern Recognition Letters*, 2003;24(14):2345–2356.

28 T.J. Ellis. Performance metrics and methods for tracking in surveillance. In *Proceedings of the Third International Workshop on Performance Evaluation of Tracking and Surveillance (PETS'2002)*, Copenhagen, June 2002, pp. 26–31.

29 Ç. Erdem, B. Sankur, and A.M. Tekalp. Metrics for performance evaluation of video object segmentation and tracking without ground-truth. In *IEEE International Conference on Image Processing (ICIP'01)*, Thessaloniki, Greece, October 2001.

30 C.J. Needham and R.D. Boyle. Performance evaluation metrics and statistics for positional tracker evaluation. In *International Conference on Computer Vision Systems (ICVS'03)*, Graz, Austria, April 2003, pp. 278–289.

31 A. Senior, A. Hampapur, Y. Tian, L. Brown, S. Pankanti, and R. Bolle. Appearance models for occlusion handling. In *Proceedings of the Second International Workshop on Performance Evaluation of Tracking and Surveillance (PETS'2001)*, Hawaii, Kauai, December 2001.

32 D. Comaniciu, V. Ramesh, and P. Meer. Real-time tracking of non-rigid objects using mean shift. In *IEEE Conference on Computer Vision and Pattern Recognition (CVPR'00)*, Hilton Head, South Carolina, June 2000, pp. 2142–2151.

33 K. Nummiaro, E. Koller-Meier, and L. Van Gool. Color features for tracking non-rigid objects. *Chinese Journal of Automation*, 2003;29(3):345–355. Special issue on visual surveillance.

34 J.M. Martinez, R. Koenen, and F. Pereira. MPEG-7 the generic multimedia content description standard, part 1. *IEEE Multimedia*, 2002;9(2):78–87.

Chapter 4

A distributed domotic surveillance system

R. Cucchiara, C. Grana, A. Prati and R. Vezzani

4.1 Introduction

In the houses of today and tomorrow the presence of capillary distributed computer systems is going to be more and more intense. In recent years the number of commercial products for the 'intelligent' house has grown considerably. The term 'domotics' (from the Latin word *domus*, which means 'home', and informatics) groups all those solutions that aim at automating certain processes for everyday life in the house. These solutions address some important issues related to domestic life: security and safety (anti-intruder alarms, fire-watching systems, gas leak detection, remote video surveillance); access control (CCTV-based video control or video entry phone); appliance control (remote programming and monitoring, diagnostics, maintenance); energy control and saving, comfort (heating and air conditioning control, light control, UPS management); actuations and motorisations (shutters and frames' control, gate control).

A new emerging term in this field is 'homevolution'; that includes both the standard domotic solutions and the new advanced solutions exploiting information technology, such as

- telecommunication: telephony, wireless communication, radio communication;
- informatics and telematics: home computing, home-office, telework;
- commerce, banking, tourism, etc.: e-commerce, home banking, e-tourism, distance learning;
- information and entertainment: video-TV, DVD, digital TV, home theatre, video games;

- intelligent control: voice-activated commands, computer vision-based surveillance, biometric authentication;
- health and well-being: telemedicine, teleassistance, telediagnosis.

These last two classes of applications are of particular interest for the social benefits they can bring and will be the main topic of this chapter. In-house safety is a key problem because deaths or injuries due to domestic incidents grow each year. This is particularly true for people with limited autonomy, such as the visually impaired, elderly or disabled. Currently, most of these people are aided by either a one-to-one assistance with an operator or an active alarm system in which a button is pressed by the person in case of an emergency. Also, continuous monitoring of the state of the person by an operator is provided by using televiewing by means of video surveillance or monitoring worn sensors.

However, these solutions are still not fully satisfactory, since they are too expensive or require an explicit action by the user (e.g. to press a button), which is not always possible in emergency conditions. Furthermore, in order to allow ubiquitous coverage of the person's movements, indoor environments often require a distributed setup of sensors, increasing costs and/or the required level of attention (since, for example, the operator must look at different monitors). To improve efficiency and reduce costs, on the one hand the hardware used must be as cheap as possible and, on the other hand, the system should ideally be fully automated to avoid both the use of human operators and explicit sensors.

Standard CCTV surveillance systems are not so widespread in domestic applications since people do not like to be continuously controlled by an operator. Privacy issues and the 'big brother' syndrome prevent their capillary distribution, even if the technology is now cheaper (cameras, storage and so on). Conversely, new solutions, fully automated and without the need for continuous monitoring by human operators, are not invasive and can be acceptable in a domotic system.

Automating the detection of significant events by means of cameras requires the use of computer vision techniques able to extract objects from the scene, characterise them and their behaviour and detect the occurrence of significant events. People's safety at home can be monitored by computer vision systems that, using a single static camera for each room, detect human presence, track people's motion, interpret behaviour (e.g. recognising postures), assess dangerous situations completely automatically and allow efficient on-demand remote connection. Since the features required by these systems are always evolving, general purpose techniques are preferable to ad hoc solutions.

Although most of the problems of visual surveillance systems are common to all the contexts, outdoor and indoor surveillance have some different requirements and, among indoor scenarios, in-house surveillance has some significant peculiarities. Indoor surveillance can, in a sense, be considered less complex than outdoor surveillance: the scene is known *a priori*, cameras are normally fixed and are not subject to vibration, weather conditions do not affect the scene and the moving targets are normally limited to people. Despite these simplified conditions, in-house scenes are

Table 4.1   *Summary of in-house surveillance problems, with their cause and effect,*
and the solutions we propose

Problem	Cause	Effect	Solution
Large shadows	Diffuse and close sources of illumination	Shape distortion and object merging	Shadow detection in the HSV colour space (Section 4.2)
Object displacement	Non-static scene	Reference background changes	Statistical and knowledge-based background model (Section 4.2)
Track-based occlusions	Position of the camera; interaction between humans	Tracking problems; shape-based algorithms are misled	Probabilistic tracking based on appearance models (Section 4.3)
Object-based occlusion	Presence of many objects		
Deformable object model	Non-rigid human body	Shape-based algorithms are misled	Probabilistic tracking based on appearance models (Section 4.3)
Scale-dependent shapes	Position and distance of the camera	Size of people depends on their distance from the camera	Camera calibration and body model scaling (Section 4.4)
Full coverage of people's movements	Non-opened, complex environments	Impossibility of covering all the movements with a single view	Camera handoff management (Section 4.5)
Accessibility	Need for every time/ everywhere access and large bandwidth	Inaccessibility for devices of limited capability	Content-based video adaptation (Section 4.6)

characterised by other problems that increase the difficulties of surveillance systems
(see Table 4.1):

1. Multi-source artificial illumination in the rooms affects moving object detection
    due to the generation of large shadows connected to the objects. Shadows affect
    shape-based posture classification and often lead to object under-segmentation.
    Thus, reliable shadow detection and removal is necessary.

2. Houses are full of different objects that are likely to be moved. The object segmentation process based on background suppression (usual for fixed camera systems) must react as quickly as possible to the changes in the background. This calls for 'knowledge-based' background modelling.
3. The main targets of tracking algorithms are people with sudden or quick appearance and shape changes, self-occlusions and track-based occlusions (i.e. occlusion between two or more moving people); moreover, tracking must deal with large background object-based occlusions due to the presence of objects in the scene.
4. Due to the presence of numerous objects in the scene, the posture classification algorithm must cope with partial (or total, but short-term) body occlusions. Since the posture of a person changes slowly, a method of assuring the temporal coherence of the posture classification should be used.
5. As mentioned above, house environments are often very complex. It is almost impossible to cover all the movements of a person within a house with a single camera. To ensure control in every place of the house, a distributed multi-camera system must be employed. Effective techniques for handling camera handoff and for exploiting distributed information must be implemented.
6. To ensure reliable and timely assistance in dangerous situations, all the time and from everywhere, remote access to the cameras must be granted from any type of device. Moreover, privacy issues should be addressed by transmitting information only when necessary. Thus, a content-based video adaptation module must be implemented to also allow access to devices with limited capabilities (such as PDAs or UMTS cellular phones).

This chapter of the book reports on our efforts at addressing all these issues with an approach as general (i.e. application- and context-independent) as possible. The next sections will discuss current solutions to these issues with reference to the current literature. Points 1 and 2 (shadow suppression and efficient background modelling) require a precise object segmentation: the proposed approach is based on adaptive and selective background update and on the detection of shadows in the HSV colour space (Section 4.2). The management of occlusions and deformability of the human body model (point 3) are addressed using a probabilistic tracking algorithm based on the appearance and the shape of the object. The result is an occlusion-proof tracking algorithm fully described in Section 4.3. This tracking has the unique characteristic of providing track features that are insensitive to human body deformation and thus it is suitable for a correct posture classification method (point 4) based on the object's shape and temporal coherence (Section 4.4). How to cope with a distributed multi-camera system (point 5) is discussed in Section 4.5, in which a novel approach for warping the appearance and the shape of the object between two consecutive views is proposed. Section 4.6 briefly reports the proposed solution for allowing universal access to the multi-camera system by means of content-based video adaptation (point 6). Finally Section 4.7 gives an overall picture, proposing an integrated system used for domotic applications, and presents conclusions.

4.2   People segmentation for in-house surveillance

Background suppression-based algorithms are the most common techniques adopted in video surveillance for detecting moving objects. In indoor surveillance the problem is strictly related to the distinction between real objects and shadows. The problems of shadow detection and precise background modelling have been thoroughly addressed in the past, proposing statistical and adaptive background models [1–4] and colour-based shadow-removing algorithms [1,2,5,6]. For instance, in Reference 3 motion segmentation is based on an adaptive background subtraction method that models each pixel as a mixture of Gaussians and uses an online approximation to update the model. Each time the parameters of the Gaussians are updated, the Gaussians are evaluated using a simple heuristic to hypothesise which are most likely to be part of the 'background process'. This yields a stable, real-time outdoor tracker that reliably deals with relatively fast illuminaton changes, repetitive motions from clutter (tree leaves motion) and long-term scene changes. Other more complex proposals make use of multi-varied Gaussians whose variables are the pixel's values in the previous N frames, such as in Reference 2. In Reference 6, Stauder *et al.* proposed to compute the ratio of the luminance between the current frame and the previous frame. A point is marked as shadow if the local variance (in a neighbourhood of the point) of this ratio is small, i.e. the variance is the same on the whole shadow region. Discriminant criteria are applied to validate the regions detected: for example, they must have an image difference larger than camera noise and should not contain any moving edges.

A peculiarity of in-house surveillance is that segmentation must not be limited to generic moving objects but also extended to stationary objects classified as people. In other words, salient objects are both generic moving objects and stationary people, in order to correctly analyse people's behaviour and avoid missing dangerous situations, such as falls. Therefore, the aim of the segmentation process is twofold: first, the detection of salient objects with high accuracy, limiting false negatives (pixels of objects that are not detected) as much as possible; second, the extraction of salient objects classified as people also whenever they are not in motion. Thus, the background model should be adaptive with both high responsiveness, avoiding the detection of transient spurious objects, such as cast shadows, static non-relevant objects or noise, and high selectivity in order to not include in the background the objects of interest.

A robust and efficient approach is to detect visual objects by means of background suppression with pixel-wise temporal median statistics, that is, a statistical approach to decide whether the current pixel value is representative of the background information is employed. In this way, a background model for non-object pixels is constructed from the set:

$$S = \{I^t, I^{t-\Delta t}, \ldots, I^{t-(n-1)\Delta t}\} \cup \left\{ \underbrace{B^t, \ldots, B^t}_{w_b \text{ times}} \right\}. \qquad (4.1)$$

The set contains n frames sub-sampled from the original sequence taking one frame every Δt (for indoor surveillance this value can be set to 10) and an adaptive factor that is obtained by including the previous background model w_b times. The new value for the background model is computed by taking the median of the set S, for each pixel p:

$$B_{stat}^{t+\Delta t}(p) = \arg \min_{x \in S(p)} \sum_{y \in S(p)} \max_{c=R,G,B} (|x_c - y_c|). \qquad (4.2)$$

The distance between two pixels is computed as the maximum absolute distance between the three channels R, G, B, which has experimentally proved to be more sensitive to small differences and is quite consistent in its detection ability.

Although the median function has been experimentally proven as a good compromise between efficiency and reactivity to luminance and background changes (e.g. a chair that is moved), the background model should not include the interesting moving objects if their motion is low or zero for a short period (e.g. a person watching the TV). However, precise segmentation indoors should exploit the knowledge of the type of the object associated with each pixel to selectively update the reference background model. In order to define a general-purpose approach, we have defined a framework where visual objects are classified into three classes: actual visual objects, shadows or 'ghosts', that is, apparent moving objects typically associated with the 'aura' that an object that begins moving leaves behind itself [1].

Figure 4.1 shows an example in which all these types appear. In this sequence, a person moves into the scene and opens the cabinet's door (Figure 4.1(a)). In the background model (Figure 4.1(b)) the cabinet's door is still closed and this results in the creation of a ghost (Figure 4.1(c)) and a corresponding 'ghost shadow' (very unusual) due to the minor reflection of the light source onto the floor. The moving person is classified as moving visual object (MVO) and the connected shadow as 'MVO shadow'.

Moving shadows and objects are not included in the background, while ghosts are forced into the background to improve responsiveness. To distinguish between apparent and actual moving objects, a post-processing validation step is performed

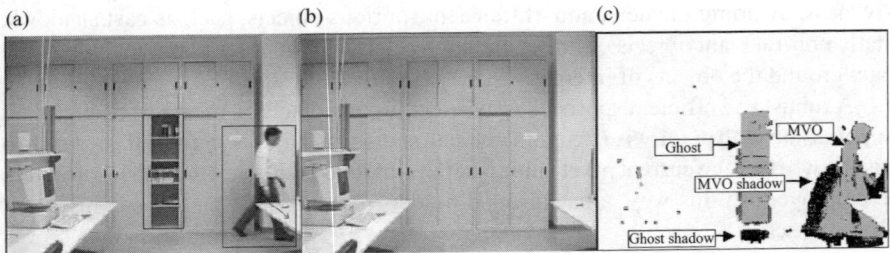

Figure 4.1   *Examples of pixel classification and selective background update: (a) shows the current frame and the resulting moving objects, (b) the current background model (note that in the background the cabinet's door is still closed) and (c) the resulting foreground pixels and their classification*

by verifying the real motion of the object. This verification can be simply done by counting the number of object's pixels changed in the last two frames.

Moreover, in the case of surveillance of people, the background updating process must take into account whether or not the detected object is a person: if it is classified as a person, it is not used to update the background, even if it is stationary, to avoid the person being included in the background and its track being lost. The classification of objects into people is often based on both geometrical and colour features [7,8]. Without the need for a precise body model, people are distinguishable from other moving objects in the house through a geometrical property checking (ratio between width and height of the bounding box) and by verifying the presence of elliptical connected areas (assumed to be the person's face).

The presence of shadows is a critical problem in all video surveillance systems, and a large effort has been devoted to shadow detection [9]. This is even more crucial in indoor, in-house surveillance where both artificial and natural light sources interact. In fact, in indoor environments many shadows are visible and no straightforward assumptions on the direction of the lights can be made. Typically, shadows are detected by assuming that they darken the underlying background but do not significantly change its colour. Thus, a model of shadow detection in the HSV colour space is proposed as in the following equation:

$$SP(p) = \begin{cases} 1 & \text{if } \alpha \leq \dfrac{I_V(p)}{B_V(p)} \leq \beta \wedge |I_S(p) - B_S(p)| \leq \tau_S \wedge D(p) \leq \tau_H \\ 0 & \text{otherwise} \end{cases},$$

(4.3)

where I is the current frame, B is the current background model and the subscripts H, S and V indicate the hue, saturation and value components, respectively. The distance $D(p)$ is computed as

$$D(p) = \min(|I_H(p) - B_H(p)|, 2\pi - |I_H(p) - B_H(p)|).$$

(4.4)

Most shadow detection algorithms exploit a certain number of parameters to obtain the best trade-off between shadow detection and foreground preservation and to adapt to illumination changes. Ideally, a good shadow detector should avoid using parameters or, at least, only use a fixed and stable set. Using the algorithm described here, the results are acceptable outdoors or when the camera acquisition system is good enough (as in the case of Figure 4.2(a) and (b), where white pixels indicate shadow points). Instead, when a low-quality camera (Figure 4.2(c)) is used (as is typical in the case of an in-house installation), the image quality can prevent us from detecting a good shadow. In this case, the chosen shadow parameters (reported in Figure 4.2(g)) can lead to over-segmentation (Figure 4.2(d)) or under-segmentation (Figure 4.2(f)), and correct segmentation can be hard to achieve.

Our experiments demonstrate that under-segmentation is not acceptable since the shape distorted by cast shadows (as in Figure 4.2(f)) cannot be easily classified as a person, and people walking in a room are easily merged in a single visual

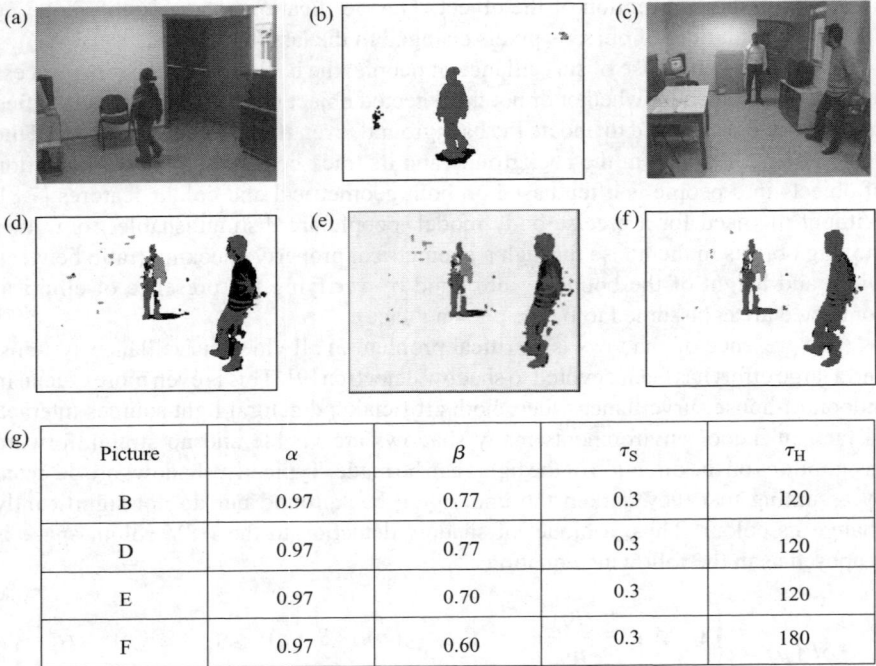

Figure 4.2   *Examples of shadow detection under different conditions. (a) and (c)
are the input frames, while (b), (d), (e) and (f) are the outputs of the
shadow detector: white pixels correspond to shadows. (g) The shadow
detection parameters adopted*

object. Conversely, the parameters of shadow detection can be set to accept a possible over-segmentation (as in Figure 4.2(e), where part of the sweater of the person on the left is detected as shadow), devolving the task of managing temporary split components or holes to the tracking module.

4.3   People tracking handling occlusions

Human motion capture and people tracking are among the most widely explored topics in computer vision. There are many reference surveys in the field: the works of Cedras and Shah [10], of Gavrila [11], Aggarwal and Cai [12] and Moeslund and Granum [13], or more recently the work by Hu *et al.* in video surveillance [14] and the work by Wang *et al.* [15]. In people tracking, in order to cope with non-rigid body motion, frequent shape changes and self-occlusions, probabilistic and appearance-based tracking techniques are commonly proposed [5,16]. In non-trivial situations, when more persons interact overlapping each other, most of the basic techniques tend to lose the previously computed tracks, detecting instead the presence of a group of people and possibly restoring the situation after the group has split up [5]. These methods aim at keeping the track history consistent before and after the occlusion only. Consequently, during an occlusion, no information about the appearance of the single person is available, limiting the efficacy of this solution in many cases. Conversely, a more challenging solution is to try to separate the group of people into individuals during the occlusion also. To this aim, the papers in References 17 and 18 propose the use of appearance-based tracking techniques.

In general, visual surveillance systems typically have the problem of coping with occlusions. In fact, occlusions can strongly affect the performance of the tracking algorithm because during them no information (or, worse, erroneous information) is available and the temporal continuity of the tracking can be violated. Different types of occlusion must be handled in different ways. In the case of track-based occlusions, that is, occlusions between visual objects (as in the case of overlapping persons), the problem is to assign shared points to the correct track. Instead, in the case of background object-based occlusions, that is, occlusions due to still objects in the background, the track model must not be updated in the occluded part in order to avoid an incorrect model update. Moreover, there is another type of shape change that could be classified as an apparent occlusion, which is mainly due to the deformability of the human body model, and to the changes of its posture. In this case the system should be able to adapt the model of the track as soon as possible.

In this chapter, we describe an approach of appearance-based probabilistic tracking, specifically conceived for handling all types of occlusions [17]. Let us assume we have, for each frame t, a set V^t of visual objects: $V^t = \{VO_1^t, \ldots, VO_n^t\}$, $VO_j^t = \{BB_j, M_j, I_j, c_j\}$. Each visual object VO_j^t is a set of connected points detected as moving by the segmentation algorithm (or, more generally, as different from the background model) and described with a set of features: the bounding box BB_j, the blob mask M_j, the visual object's colour template I_j and the centroid c_j. During the tracking execution, we compute a set τ^t of tracks at each frame t, that represents the knowledge of the objects present in the scene, $\tau^t = \{T_1^t, \ldots, T_m^t\}$ with $T_k^t = \{BB_k, AMM_k, PM_k, PNO_k, c_k, e_k\}$, where AMM_k is the appearance memory model, that is, the estimated aspect (in RGB space) of the track's points: each value $AMM_k(\mathbf{x})$ represents the 'memory' of the point appearance, as it has been seen up to now; PM_k is the probability mask; each value of $PM_k(\mathbf{x})$ defines the probability that the point $\mathbf{x}$ belongs to the track T_k; PNO_k is the probability of non-occlusion associated with the whole track, that is the probability that the track k is not occluded by other tracks; e_k is the estimated motion vector.

Typically, tracking is composed of two steps: the correspondence between the model (track) and the observation (the visual objects), and, then, the model update. If the algorithm has to deal with occlusions and shape changes, the approach needs to add more steps: VO-to-track mapping, track position refinement, pixel-to-track assignment, occlusion classification, track update.

The first three steps cope with the possible presence of more tracks in the neighbourhood of a single object or a group of segmented objects, so that the one-to-one correspondence is not always suitable. In order to integrate in a single structure all the possible conditions (VO merging, splitting and overlapping), the concept of macro-object (MO) is introduced: an MO is the union of the VOs potentially associated to the same track or set of tracks. The process of creating an MO begins with the construction of a Boolean correspondence matrix C between the sets V^t and T^{t-1}. The element $C_{k,j}$ is set to 1 if the VO_j^t can be associated to the track T_k^{t-1}. The association is established if the track (shifted into its estimated position by means of the vector $\mathbf{e}_k$) and the VO can be roughly overlapped, or, in other words, if they have a small distance between their closest points. Indoors it is very frequent that more VOs (that could be different persons or also a single person split in more VOs due to segmentation problems) are associated with one or more tracks.

Instead of evaluating the VO-to-track correspondence, an MO-to-track correspondence is more effective since it takes into account all possible errors due to visual overlaps and segmentation limitations. After having merged distinct VOs in a single MO, the tracking iterates the designed algorithm at each frame and for each pair $(MO, \tilde{\tau}^t)$. At each iteration, for each track $T_i \in \tilde{\tau}^t$, the algorithm first predicts the track's position (using previous position and motion information), then it precisely aligns the track and assigns each pixel of the MO to the track with the highest probability to have generated it. Then, the remaining parts of the tracks that are not in correspondence with detected points in the frame are classified as potential occlusions and a suitable track model update is performed.

In the track alignment phase the new position of each track k is estimated with the displacement $\delta = (\delta_x, \delta_y)$ that maximises an average likelihood function $F_k(\delta)$.

$$F_k(\delta) = \frac{\sum_{x \in MO} p(\mathbf{I}(\mathbf{x} - \delta)|\mathbf{AMM}_k(\mathbf{x})) \cdot PM_k(\mathbf{x})}{\sum_{x \in MO} PM_k(\mathbf{x})}, \tag{4.5}$$

where $p(\mathbf{I}(\mathbf{x} - \delta)|\mathbf{AMM}_k(\mathbf{x}))$ is the conditional probability density of an image pixel, given the appearance memory model. We assume this density to have a Gaussian distribution centred on the RGB value of pixel x in the appearance memory model, with a fixed covariance matrix. We also suppose that the R,G,B variables are uncorrelated and with identical variance. This function will have higher values when the appearance of the object shows little or no changes over time. The δ displacement is initialised with the value $\mathbf{e}_k$ and searched with a gradient descent approach. The alignment is given by $\tilde{\delta}_k = \arg \max_\delta (F_k(\delta))$.

We also calculate another value, confidence, that is the percentage of track points, weighted with their probability, that are visible on the current frame and belong to the MO:

$$confidence_k = \frac{\sum_{x \in MO} PM_k(\mathbf{x})}{\sum_{x \in T_k} PM_k(\mathbf{x})}. \tag{4.6}$$

In case the product of likelihood and confidence is too low, meaning that the track is occluded and its visible part does not have a good match with the appearance model, the displacement $\tilde{\delta}$ found is not reliable, and the previously calculated motion vector $\mathbf{e}_k$ is used as the most reliable displacement. The alignment is iterated for all the tracks T_k associated with an MO, with an order proportional to their probability of non-occlusion PNO_k. After each fitting computation, the points of the MO matching a track point are removed and not considered for the following tracks' fitting.

Once all the tracks have been aligned, each point of the MO is independently assigned to a track. Making the alignment phase independent from the assignment phase gives some advantages. If we assign the pixels only on the basis of their correspondence to the tracks' position and depth estimation, two problems could arise: the alignment is at the region level, so if the object has changed its shape during an occlusion, many pixels could be wrongly assigned. Furthermore, if the tracks' depth estimation is incorrect, recovery from errors could be hard, because each pixel corresponding to the rearmost track mask would be assigned to it, while they actually belong to the foremost track.

Due to occlusions or shape changes, after the assignment some points of the tracks remain without any correspondence with an MO point. Other techniques that exploit probabilistic appearance models without coping with occlusions explicitly use only a set of assigned points $A_k \subseteq$ MO to guide the update process [16]: for each pixel $\mathbf{x} \in T_k$, the probability mask is reinforced if $\mathbf{x} \in A_k$, otherwise it is decreased.

In this scheme, the adaptive update function is enriched by the knowledge of occlusion regions. The set of non-visible points $NV_k^t = \{x \in T_k^t \vee x \notin A_k\}$ contains the candidate points for occlusion regions: in general, they are the tracks' points that are no longer visible at the frame t. After a labelling step, a set of non-visible regions (of connected points of NV_k^t) is created; sparse points or too small regions are pruned and a set of candidate occlusion regions $\{COR_{ki}^t\}_{i=1...r_k}$ is created. As discussed above, non-visible regions can be classified into three classes: track-based occlusions R_{TO}, background object-based occlusions R_{BOO} and apparent occlusions R_{AO}.

The presence of occlusions can be detected with the confidence value of Equation (4.6) decreasing below an alerting value, since in the case of occlusion the track's shape changes considerably. The track model in the points of actual occlusions (R_{TO} and R_{BOO}) should not be updated since the memory of the people appearance should not be lost; conversely, if the confidence decreases due to a sudden shape motion (apparent occlusion), not updating the track would create an error. The solution is a selective update according to the region classification. In particular, $\forall \mathbf{x} \in T^t$

$$\mathbf{PM}^t(\mathbf{x}) = \begin{cases} \lambda \mathbf{PM}^{t-1}(\mathbf{x}) + (1-\lambda) & \mathbf{x} \in A_K, \\ \mathbf{PM}^{t-1}(\mathbf{x}) & (\mathbf{x} \in R_{TO}) \vee (\mathbf{x} \in R_{BBO}), \\ \lambda \mathbf{PM}^{t-1}(\mathbf{x}) & \text{otherwise.} \end{cases} \tag{4.7}$$

$$\mathbf{AMM}^t(\mathbf{x}) = \begin{cases} \lambda \mathbf{AMM}^{t-1}(\mathbf{x}) + (1-\lambda)\mathbf{I}^t(\mathbf{x}) & \mathbf{x} \in A_K \\ \mathbf{AMM}^{t-1}(\mathbf{x}) & \text{otherwise.} \end{cases} \tag{4.8}$$

When the track is generated, $\mathbf{PM}^t(x)$ is initialised to an intermediate value (equal to 0.4 when $\lambda = 0.9$), while the appearance memory model is initialised to the image $\mathbf{I}^t(\mathbf{x})$. To compute the probability of non-occlusion, $P(T_k) \equiv PNO^t(T_k)$, an adaptive filter that considers the number of contended points is employed [17].

The detection of the track-based occlusion is straightforward, because the position of each track is known, and it is easy to detect when two or more tracks overlap.

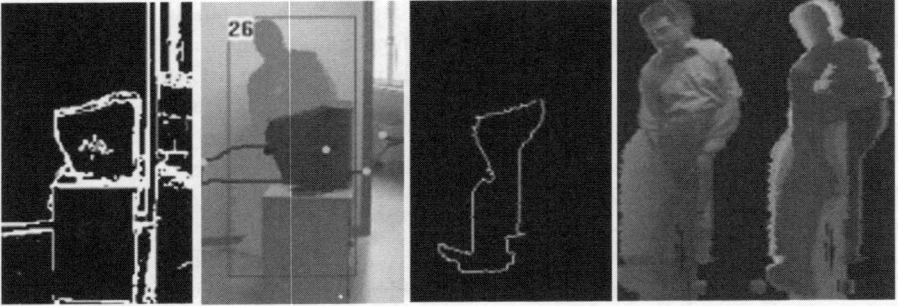

Figure 4.3   Edge pixel classification

R_{TO} regions are composed by the points shared between track T_k and other tracks T_i but not assigned to T_k.

To distinguish between the second type (R_{BOO}) and the third type (R_{AO}) is more challenging, because in a general case the position and the shape of the objects in the background are not known. However, it is possible to employ the background edge's image, which is an image that contains high colour variation points, among which the edges of the objects can be easily detected. The edge pixels that touch visible pixels of the track are classified as bounding pixels while the others are said to be not-bounding pixels. If the majority of the bounding pixels match background edges, we can suppose that an object hides a part of the track, and the region is labelled as R_{BOO}, otherwise as R_{AO}. In Figure 4.3 an illustrative example is shown: a person is partially occluded by furniture that is included in the background image. As a consequence, part of its body is not segmented. The reported images show (from left to right) the edge map of the background image, the segmented VO, the edges of the COR with the bounding pixels matching the background edges, the appearance memory model and the probability mask.

In conclusion, this tracking algorithm provides, at each frame, the silhouette and the appearance memory model of the segmented people, even in the case of difficult situations, such as large and long-lasting occlusions.

4.4   Posture classification in deformable models with temporal coherence

In the case of health care in domestic environments, most of the relevant situations can be inferred by analysing the posture of the monitored person. In fact, posture can be useful in understanding the behaviour of the person.

Posture classification systems proposed in the past can be differentiated by the more or less extensive use of a 2D or 3D model of the human body [13]. In accordance with this, we can classify most of them into two basic approaches. From one perspective, some systems (such as that proposed in Reference 7) use a direct approach and base the analysis on a detailed human body model. In many of these cases, an incremental predict–update method is used, retrieving information from every body part. Many systems use complex 3D models and require special and expensive equipment, such as 3D trinocular systems [19], 3D laser scanners [20], thermal cameras [21] or multiple video cameras to extract 3D voxels [22]. However, for in-house surveillance systems, low-cost solutions are preferable, and thus stereo-vision or 3D multi-camera systems have to be discarded. In addition, these are often too constrained to the human body model, resulting in unreliable behaviours in the case of occlusions and perspective distortion that are very common in cluttered, rela-tively small, environments like a room. Consequently, algorithms of people's posture classification should be designed to work with simple visual features from a sin-gle view, such as silhouettes, edges and so on [7,13,23]. Most of the approaches in the literature extract a minimal set of low-level features exploited in more or less sophisticated classifiers, such as neural networks [24]. However, the use of NNs presents several drawbacks due to scale dependence and unreliability in the case of occlusions. Another large class of approaches is based on human silhouette analy-sis. The work of Fujiyoshi *et al.* [25] uses a synthetic representation (star skeleton) composed of outmost boundary points. The approach in Reference 7 exploits the projection histograms as sufficient features to describe the posture.

In the approach proposed here, posture classification is carried out after a machine learning process to learn the visual models corresponding to different postures; then, a probabilistic classification is provided for recognising the trained postures. In in-house surveillance, the system could be limited to classifying main postures only but it should be able to recognise abrupt changes in the posture that could infer a dangerous situation. For instance, the main postures could be standing, crouching, sitting and lying.

A cost-effective and reliable solution is based on intrinsic characteristics of the silhouette only, to recall the person's posture, exploiting projection histograms $PH = (\vartheta(x); \pi(y))$ that describe the way in which the silhouette's shape is projected on the x and y axes. Examples of projection histograms are depicted in Figure 4.4.

Although projection histograms are very simple features, they have proven to be sufficiently informative to discriminate between the postures we are interested in. However, these descriptors have two limitations: they depend on the silhouette's size (due to the perspective, the size of the detected VO can be different) and they are too sensitive to the unavoidable non-rigid movements of the human body. To mitigate the first point, the support point for each person (i.e. the contact point between the person and the floor) is calculated and, through calibration information, its distance from the camera is computed to scale the silhouettes of the person proportionally with it. A detailed description of the support point estimation and tracking algorithm is reported in the next subsection.

For the second problem, a suitable model capable of generalising the peculiarities of a training set of postures has been adopted. In particular, a pair of projection prob-abilistic maps (PPMs, $\Theta_i(x, y)$ and $\Pi_i(x, y)$) for each posture are computed through a supervised machine-learning phase. By collecting a set of T_i manually classified

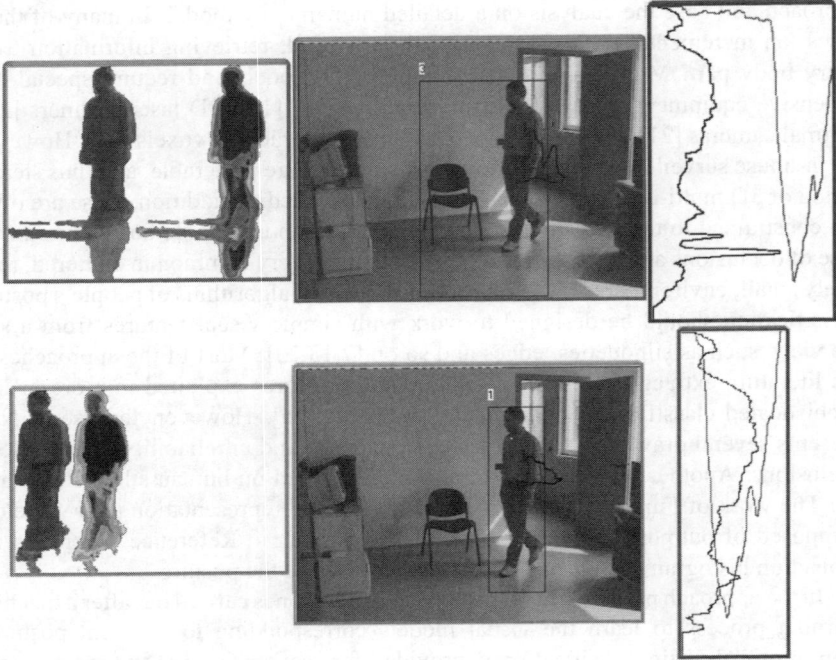

Figure 4.4   Tracks, VO extent and projection histograms computed without (top) or with shadow suppression (bottom)

silhouettes for each posture i, the probabilistic approach included in the PPMs allows us to filter irrelevant moving parts of the body (such as the arms and the legs) that can generate misclassifications of the posture:

$$\Theta_i(x,u) = \frac{1}{T_i} \cdot \sum_{t=1}^{T_i} g(\theta_i^t(x), u)$$

$$\Pi_i(v,y) = \frac{1}{T_i} \cdot \sum_{t=1}^{T_i} g(v, \pi_i^t(y)) \tag{4.9}$$

with

$$g(s,t) = \frac{1}{|s-t|+1}. \tag{4.10}$$

These maps can be considered as a representation of the conditional probability function to have a projection histogram given the posture i:

$$\Theta_i(x,u) \propto P(\theta(x) = u | \text{posture} = i). \tag{4.11}$$

Figure 4.5   Appearance images (left) and probability masks (right) of a person's track in different postures

At the classification stage, the projection histograms PH computed over a blob are compared with the PPM of each class, obtaining the following probabilities:

$$P(\text{PH}|\text{posture} = i) = P(\hat{\theta}|\text{posture} = i) \cdot P(\hat{\pi}|\text{posture} = i) \quad \forall i. \qquad (4.12)$$

Supposing the priors of the posture classes are uniformly distributed, the classifier selects the posture that maximises the reported conditional probabilities.

If the lower-level segmentation module produces correct silhouettes, then this classifier is precise enough. However, by exploiting knowledge embedded in the tracking phase, many possible classification errors due to the imprecision of the blob extraction can also be corrected. In particular, the projection histograms are computed on the appearance memory models of the tracks instead of on the blobs (or VOs) extracted during the background suppression stage. Figure 4.5 shows some examples of the tracks appearance memory model for different postures over which the projection histograms are computed.

4.4.1   Support point tracking

The correct estimation of the SP (support point) position during an occlusion also is a key requirement for correct people scaling. The SP is normally coincident with the feet's position, but it could differ when the person is lying down on the floor. If the camera has no roll angle and the pan angle is low enough, the support point can be estimated taking the maximum y coordinate and the average of the x coordinates of the lowest part of the blob (Figure 4.6).

This simple algorithm fails in the presence of occlusions: if the occlusion includes the feet of the person, in fact, the y coordinate of the SP becomes not valid. The SP cannot be computed either on the blob or on the probability mask: the first because the blob is incomplete and the second because the computation is imprecise and unreliable, especially in body parts with frequent motion such as the legs. To solve this problem, a dedicated tracking algorithm for the SP coordinates has been developed. Two independent constant-velocity Kalman filters are adopted: the first is a standard Kalman filter on the x coordinate, while the second Kalman filter for the y coordinate of SP is used in two modalities – when the person is visible the filter considers both the forecast and the measure (as usual), while when the person is occluded the

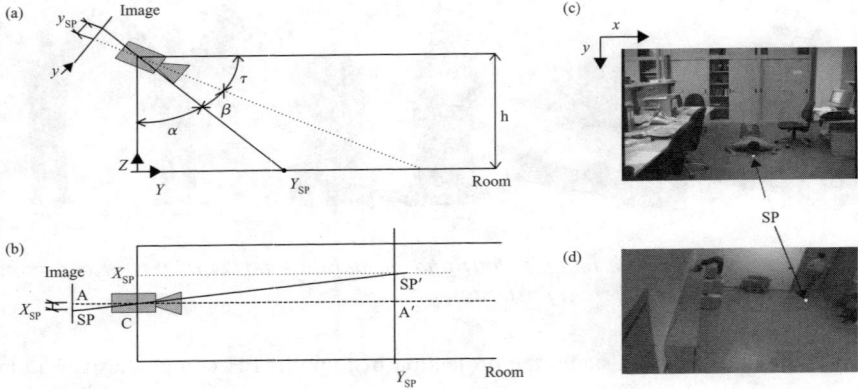

Figure 4.6 SP estimation

measure of the y coordinate is ignored and the filter exploits only the forecast to estimate the current position.

The two filters have the same parameters and exploit the constant velocity assumption. Using the well-known notation of the discrete Kalman filter,

$$x(k + 1) = \boldsymbol{\Phi} \cdot x(k) + v$$

$$(4.13)$$

$$z(k) = \mathbf{H} \cdot x(k) + w,$$

the adopted matrices are

$$x(k) = \begin{pmatrix} pos_k \\ vel_k \end{pmatrix} \quad \mathbf{H} = \begin{pmatrix} 1 & -1 \\ 1 & 0 \end{pmatrix} \quad \mathbf{Q} = \begin{pmatrix} 0 & 0 \\ 0 & \gamma \end{pmatrix}$$

$$(4.14)$$

$$z(k) = \begin{pmatrix} pos_{k-1} \\ pos_k \end{pmatrix} \quad \boldsymbol{\Phi} = \begin{pmatrix} 1 & 1 \\ 0 & 1 \end{pmatrix} \quad \mathbf{R} = \begin{pmatrix} \lambda & 0 \\ 0 & \lambda \end{pmatrix}.$$

The measurement noise covariance matrix $\mathbf{R}$ of the Gaussian variable w is computed assuming that the two measured positions could be affected by noise that is frame-by-frame independent and time-constant, while the process noise covariance $\mathbf{Q}$ of the variable v assumes that the process noise affects only the velocity terms. In our implementation, the parameters are set to $\lambda = 200$ and $\gamma = 0.5$.

The results obtained by the introduction of these two independent filters during a strong occlusion are visible in Figure 4.7.

Let the SP position and the homography of the floor plane be known; the distance between the person and the camera can be computed. Supposing that the image of the person entirely lies on a plane parallel to the camera; the blob and the tracking images can be projected into a metric space by a linear scaling procedure.

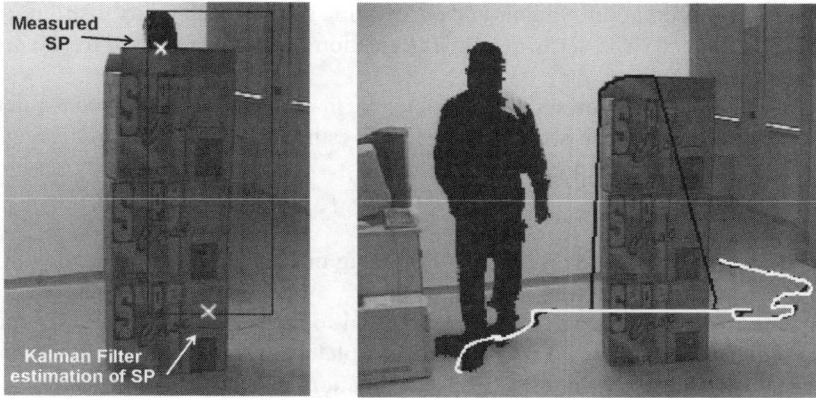

Figure 4.7   (a) SP tracking with Kalman filters; (b) trajectories of SP estimated over the blob (shown in black) and tracked with the Kalman filter (shown in white)

4.4.2   Temporal coherence posture classifier

Despite the improvements given by the use of an appearance mask instead of blobs, a frame-by-frame classification is not reliable enough. However, the temporal coherence of the posture can be exploited to improve the performance: in fact, the person's posture changes slowly, through a transition phase during which its similarity with the stored templates decreases. To this aim, a hidden Markov models formulation has been adopted. Using the notation proposed by Rabiner in his tutorial [26], we define the followings sets:

- The state set **S**, composed by four states:

$$S = \{S_1, S_2, S_3, S_4\} = \text{Main\_Postures}. \tag{4.15}$$

- The initial state probabilities $\Pi = \{\pi_i\}$: the initial probabilities are set to have the same value for each state ($\pi_i = \frac{1}{4}$). By introducing the hypothesis that a person enters a scene standing, it is possible to set the vector Π with all the elements equal to 0 except for the element corresponding to the standing state (set to the value 1). However, the choice of the values assigned to the vector Π affects the classification of the first frames only, and then it is not relevant.
- The matrix **A** of the state transition probabilities, computed as a function of a reactivity parameter α. The probabilities of remaining in the same state and passing to another state are considered equal for each posture. Then, the matrix **A** has the following structure:

$$A = A(\alpha) = \{A_{ij}\}, \quad A_{ij} = \begin{cases} \alpha & i = j, \\ \dfrac{1-\alpha}{3} & i \neq j. \end{cases} \tag{4.16}$$

In our system we use $\alpha = 0.9$.

Then, the observation symbols and observation symbol probability distribution **B** have to be defined. The set of possible projection histograms is used as the set of observation symbols.

Even if the observation set is numerable, the matrix B is not computed explicitly, but the values b_j for each state (posture) j are estimated frame-by-frame using the above defined projection probabilistic maps:

$$b_j = P_j = P(\text{PH}|\text{posture} = S_j). \tag{4.17}$$

Then, at each frame, the probability of being in each state is computed with the forward algorithm [26].

Finally, the HMM input has been modified to keep into account the visibility status of the person. In fact, if the person is completely occluded, the reliability of the posture must decrease with time. In such a situation, it is preferable to set $b_j = 1/N$ as the input of the HMM. In this manner, the state probability tends to a uniform distribution (that models the increasing uncertainty) with a delay that depends on the previous probabilities: the higher the probability of being in a state S_j, the higher the time required to lose this certainty. To manage the two situations simultaneously and to cope with the intermediate cases, (i.e., partial occlusions), a generalised formulation of the HMM input is defined:

$$b_j = P(\text{PH}|S_j) \cdot \frac{1}{1 + n_{fo}} + \frac{1}{N} \cdot \frac{n_{fo}}{1 + n_{fo}}, \tag{4.18}$$

where n_{fo} is the number of frames for which the person is occluded. If n_{fo} is zero (i.e. the person is visible), b_j is computed as in Equation (4.17), otherwise the higher the value of n_{fo}, the more it tends to a uniform distribution.

In Figure 4.8 the benefits of the introduction of the HMM framework are shown. The results are related to a video sequence in which a person passes behind a stack of boxes always in a standing position. During the occlusion (highlighted by the grey strips) the frame-by-frame classifier fails (it states that the person is lying). On the other hand, through the HMM framework, the temporal coherence of the posture is preserved, even if the classification reliability decreases during the occlusion.

4.5   Multi-camera system management

Since the entire house is not coverable with a single camera, a distributed system must be employed. This calls for the introduction of coordination modules to solve camera handoff problems. In addition, an occlusion of the monitored person could occur during its entry within a room. In such a situation, a stand-alone appearance-based tracking cannot solve the occlusion, and the higher level processes, such as the posture classification module, would fail. Several approaches have been proposed to maintain the correspondence of the same tracked object during a camera handoff. Most of these require a partially overlapping field of view [27–29]; other ones use a feature-based probabilistic framework to maintain a coherent labelling of the objects [30]. All these works aim at keeping correspondences between the same

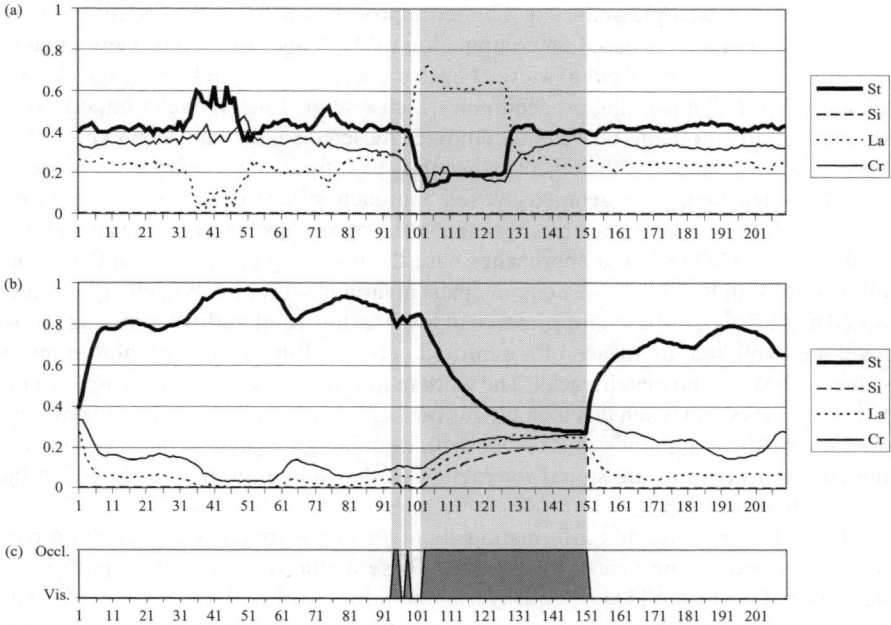

Figure 4.8   Comparison between the frame-by-frame posture classification (a) and the probabilistic classification with HMM (b) in the presence of occlusions (c)

tracked object, but none of them is capable of handling the occlusions during the camera handoff phase.

Approaches to multi-camera tracking can be generally classified into three categories: geometry-based, colour-based and hybrid approaches. The first class can be further sub-divided into calibrated and uncalibrated approaches. In Reference 31, each camera processes the scene and obtains a set of tracks. Then, regions along the epipolar lines in pairs of cameras are matched and the mid-points of the matched segments are back-projected in 3D and then, with a homography, onto the ground plane to identify possible positions of the person within a probability distribution map (filtered with a Gaussian kernel). The probability distribution maps are then combined using outlier-rejection techniques to yield a robust estimate of the 2D position of the objects, which is then used to track them. A particularly interesting method is reported in Reference 32 in which homography is exploited to solve occlusions. Single-camera processing is based on a particle filter and on probabilistic tracking based on appearance (as in Reference 16) to detect occlusions. Once an occlusion is detected, homography is used to estimate the track position in the occluded view, by using the track's last valid positions and the current position of the track in the other view (properly warped in the occluded one by means of the transformation matrix).

A very relevant example of the uncalibrated approaches is the work of Khan and Shah [30]. Their approach is based on computation of the 'edges of field of view', that is, the lines delimiting the field of view of each camera that thus define the overlapped regions. Through a learning procedure in which a single track moves from one view to another, an automatic procedure computes these edges that are then exploited to keep consistent labels on the objects when they pass from one camera to the adjacent one. In conclusion, pure geometry-based approaches have been used rarely in the literature, but they have been often mixed with approaches based on the appearance of the tracks. Colour-based approaches base the matching essentially on the colour of the tracks. In Reference 33 a colour space invariant to illumination changes is proposed and histogram-based information at low-(texture) and mid-(regions and blobs) level are exploited, by means of a modified version of the mean shift algorithm, to solve occlusions and match tracks. The work in Reference 34 uses stereoscopic vision to match tracks, but when this matching is not sufficiently reliable, colour histograms are used to solve ambiguities. To cope with illumination changes, more versions of the colour histograms are stored for each track depending on the area in which the track is detected.

Hybrid approaches mix information about the geometry and the calibration with those provided by the visual appearance. These methods are based on probabilistic information fusion [35] or on Bayesian belief networks (BBNs) [27,36,37], and sometimes a learning phase is required [38,39]. In Reference 35, multiple visual and geometrical features are evaluated in terms of their PDFs. In that paper, two models, one for the motion estimated with a Kalman filter and one for the appearance by using the polar histograms, are built and merged into a joint probability framework that is solved by using the Bayes rule. In Reference 36, geometry-based and recognition-based functioning modes are integrated by means of a BBN and evaluated with likelihood and confidence metrics. Geometrical features include epipolar geometry and homography, while recognition-based features are basically the colour and the apparent height of the track. Though very complete, this approach is very sensitive to the training of the BBN and to the calibration. A similar approach has been proposed in Reference 27, and the matching between two cameras is performed on the basis of positional and speed information mapped on 3D, thanks to the calibration and the pin-hole camera model. The MAP estimation is solved by an efficient linear programming technique. A BBN is used in Reference 37 for fusing the spatial and visual features extracted from all the cameras. The cameras are then ordered by their confidence and the two cameras with the higher value are extracted and projected on 3D to produce a set of observations of 3D space to be used as input for a tri-dimensional Kalman filter. All these solutions are more or less tailored to the specific application or to the specific camera setup. To develop a more flexible system, capable of working in many different situations, a solution proposed in the literature is to use a learning phase. In Reference 38, the colour spaces perceived by the different cameras are correlated by means of a learning phase. This process aims at building a uniform colour space for all the cameras used, together with geometrical features, for the object matching. Two different approaches, based on support vector regression and on hierarchical PCA, are compared in terms of both accuracy and computational efficiency. In Reference 39 the learning phase is used to infer the camera on which an object will appear when it exits from another camera. In addition, this phase is also used to learn how the appearance changes from one camera to another (using the colour histograms). As mentioned above, the approach used for matching the objects is often exploited (implicitly) also for solving the occlusions [31,35,36]. Some papers, however, directly address the problem of handling occlusions in multi-camera systems. For example, in Reference 34, when an occlusion is detected (i.e. the matching between the extracted track and corresponding model is not sufficiently good) the track is 'frozen' until it becomes completely visible again. In Reference 38, the geometry-based and recognition-based features provided by a non-occluded camera are exploited in the occluded view to solve the occlusion by means of a Bayesian framework and Kalman filter. In Reference 40, a geometrical method is used to obtain a probability of occlusion in a multi-camera system. In Reference 32 the track's appearance is projected through a homography from the non-occluded view to the occluded one.

In this section we describe the exploitation of the tracking algorithm in a multi-camera system, and as in Reference 32, the use of homography to transfer appearance memory models from one camera system to another.

As stated above, the probabilistic tracking is able to handle occlusions and segmentation errors in the single-camera module. However, to be robust to occlusions the strong hypothesis is made that the track has been seen for some frames without occlusions so that the appearance model is correctly initialised. This hypothesis is erroneous in a case where the track is occluded since its creation (as in Figure 4.9(b)).

Our proposal is to exploit the appearance information from another camera (where the track is not occluded) to solve this problem. If a person passes between two monitored rooms, it is possible to keep the temporal information stored into its track extracted from the first room (Figure 4.9(a)) and use it to initialise the corresponding track in the second room (Figure 4.9(b)).

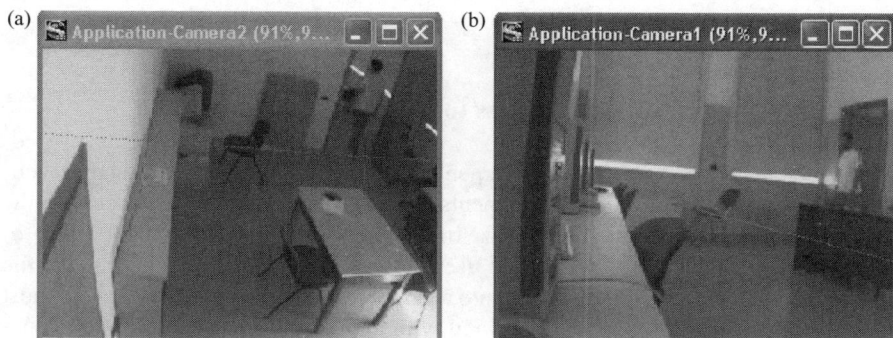

Figure 4.9   The frame extracted from the two cameras during the camera handoff. The lines are the intersections of the floor and the gate plane computed and drawn by the system

To this aim the following assumptions are used:

- The two cameras are calibrated with respect to the same coordinate system (X_W, Y_W, Z_W).
- The equation of the plane $G = f(X_W, Y_W, Z_W)$ containing the entrance is given.
- It is possible to obtain the exact instant t_{SYNC} when the person passes into the second room. To do this, the 3D position of the support point SP of the people inside the room could be used, otherwise a physical external sensor could be adopted for a more reliable trigger.
- All the track points lie on a plane P parallel to the entrance plane G and containing the support point SP (hereinafter we call P person's plane). This assumption holds if the person passes the entrance in a posture such that the variance of the points with respect to the direction normal to P is low enough (e.g. standing posture).

The first three assumptions imply only an accurate installation and calibration of cameras and sensors, while the last one is a necessary simplification to warp the track between the two points of view. Under this condition, in fact, the 3D position of each point belonging to the appearance image of the track can be computed and then its projection on a different image plane is obtained.

In particular, the process mentioned above is applied only to the four corners of the tracks and thus the homography matrix $\mathbf{H}$ that transforms each point between the two views can be computed:

$$[x_2 y_2 1]^T = \mathbf{H}_{3 \times 3} \cdot [x_1 y_1 1]^T. \tag{4.19}$$

Through $\mathbf{H}$ it is possible to re-project both the appearance image AI(x) and the probability mask PM(x) of each track from the point of view of the leaving room to the point of view of the entering one (Figure 4.10). The re-projected track is used as initialisation for the new view that can in such a manner solve the occlusion by continuing to detect the correct posture. This warping process implicitly solves the consistent labelling problem, by directly initialising the tracks in the new view.

4.6   Universal multimedia access to indoor surveillance data

As already stated, an essential requirement of automated systems for visual surveillance of people in domestic environments is that of providing visual access to the situation to remote operators in the case of emergencies. This access should be provided without the need for an explicit user request (that can be unable to make the request) and by taking privacy issues into account. Moreover, this visual access must be made possible almost everywhere and with any device (also with limited capabilities). This possibility is often called UMA (universal multimedia access) and requires the adaptation of the video content, in order to meet the device constraints and the user's requests. This process is referred to as content-based or semantic video adaptation [41].

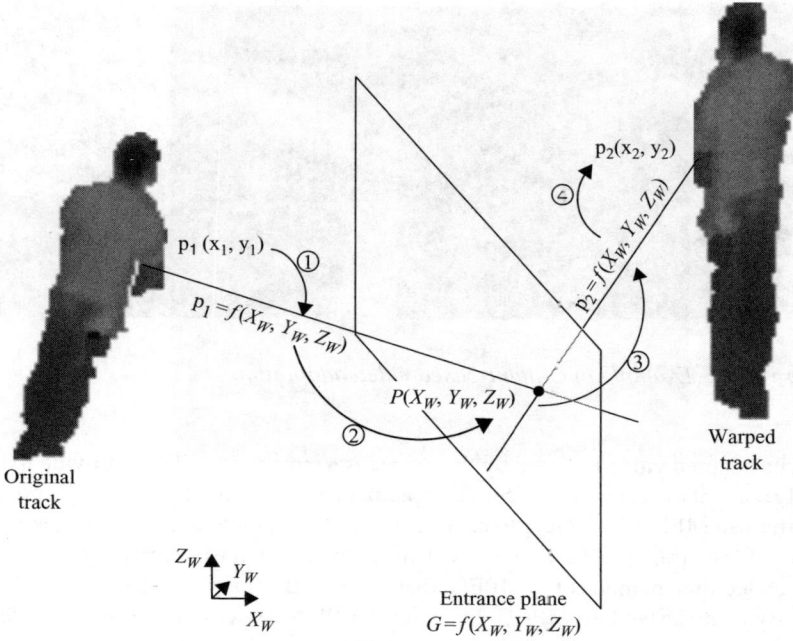

Figure 4.10 Track warping: (1) exploiting the calibration of camera 1; (2) intersection with entrance plane; (3) calibration of camera 2; (4) intersection with camera 2 plane

With these premises, the aforementioned computer vision techniques are exploited to extract the semantics from the video. This semantics drives the content-based adaptation in accordance with the user's requests. In particular, the ontology of the current application is organised into classes of relevance [41]. Each class of relevance C is defined as $C = \langle e, o \rangle$, with $e \subseteq E$ and $o \subseteq O$, where E and O are, respectively, the set of event types e_i and of object types o_i that the computer vision techniques are capable of detecting, and e and o are, respectively, sub-sets of E and O. The user can group into a class of relevance events and objects that they consider as of similar relevance, and then associate a weight to each class of relevance: the higher the weight, the more relevant the class must be considered, and less aggressive transcoding policies must be applied. In particular, limiting the transcoding process to different compression levels, the higher the weight, the less compressed the information will be and the better the quality provided.

In in-house surveillance systems, the classes with higher relevance are those including people as objects (segmented and classified by the segmentation module, and tracked by the tracking one), and falls and dangerous situations as relevant events (extracted by the behaviour analysis derived by the posture classification module). An example of content-based video adaptation is provided in Figure 4.11.

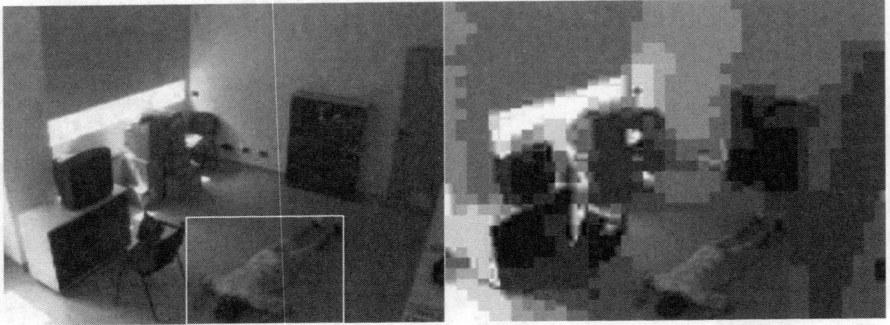

Figure 4.11   Example of content-based video adaptation

The adapted video is coded by the server as a stream, in order to provide live and timely content to the client device. To implement a coding driven by the semantics, we modified an MPEG-2 encoder to change the compression factor of the single macroblocks of each frame, in accordance with the relevance of the majority of its pixels. We have called this method SAQ-MPEG, that is, semantic adaptive quantisation-MPEG [42]. By using a standard MPEG-2 decoder for PDA devices, it is possible to decode the stream coming from the server and visualise the final adapted video on-the-fly on the PDA screen [43].

4.7   Conclusions

The techniques described in the previous sections have been implemented and integrated in a system able to access multiple cameras, extract and track people, analyse their behaviour and detect possible dangerous situations, reacting by sending an alarm to a remote PDA with which a video connection is established. The integrated system is structured as a client–server architecture, as shown in Figure 4.12. The server side contains several pipelined modules: object segmentation, tracking, posture classification and event detection. The alarms generated can be sent to a control centre or can trigger remote connections with a set of authorised users, which exploit the universal multimedia access algorithms to receive visual information of the objects (people) involved in dangerous situations [41].

According to the user's requests and the device's capabilities, the transcoding server adapts the video content, modifying the compression rate of the regions of interest, thus sending only what is required and only in the case of a significant event occurring. The transcoding policy resolver (TPR) decides which are the more appropriate policies for obtaining the best trade-off between video quality and bandwidth allocated, taking the user's preferences into account.

In the introduction, we listed some peculiarities of in-house video surveillance that must be taken into account to develop a reliable and efficient system. We can

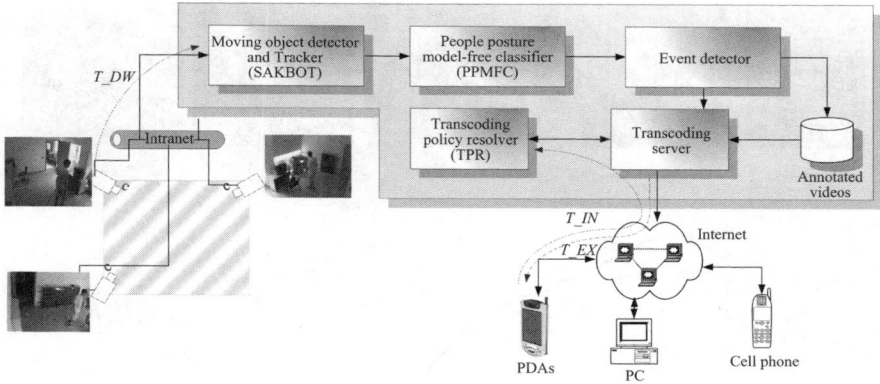

Figure 4.12   Overall scheme of the system

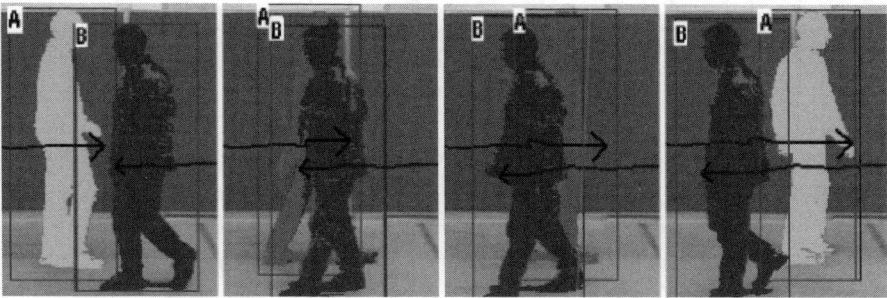

Figure 4.13   Correct track-based occlusion solving

summarise the distinctive problems to be addressed in in-house video surveillance as in Table 4.1.

The proposed solution for shadow detection proves to be reliable for detecting shadow points and discriminating them with reference to foreground points [1], and its benefits for posture classification are evident in Figure 4.4. It can be noted that the shadows cast underneath and on the left of the person heavily modify the shape and, consequently, the projection histograms. The probabilistic tracking described in Section 4.3 solves many problems due to both occlusions and non-rigid body models.

Figure 4.13 shows an example of track-based occlusion in which the tracks associated with the two people (indicated by capital letters) are never lost, even if total occlusion is observed. Figure 4.14 illustrates a case of object-based occlusion and shows how the confidence decreases when the person goes behind the object and the model is not updated.

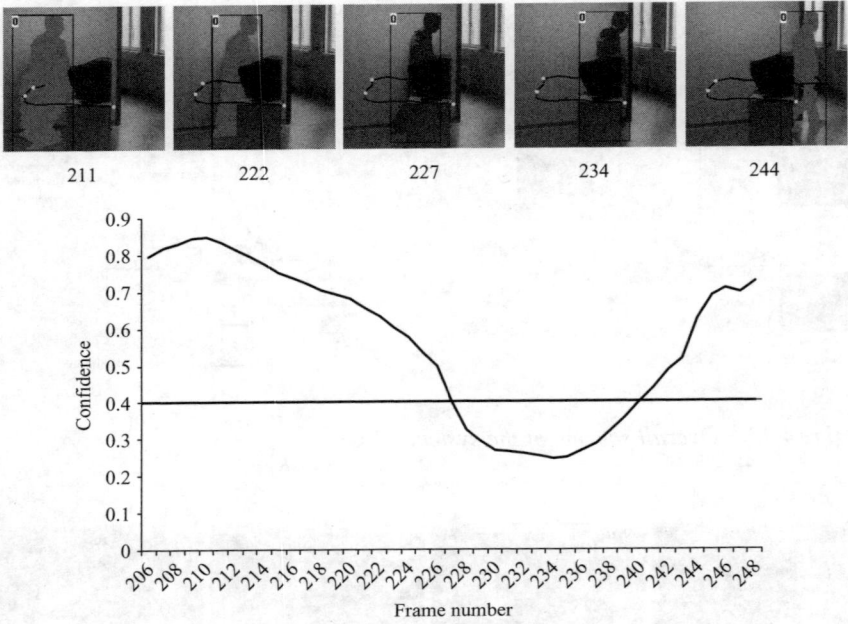

Figure 4.14   Example of object-based occlusion and the corresponding confidence value

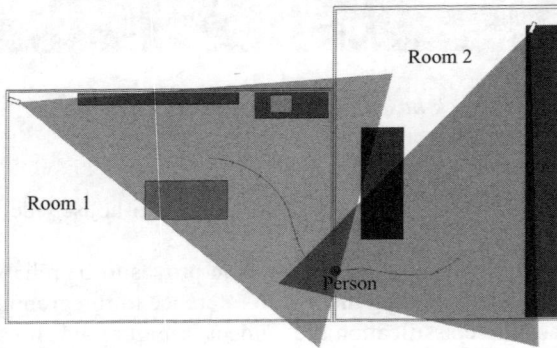

Figure 4.15   A scheme of the two rooms used for our tests

The way in which the non-rigid body model is treated with the appearance model and probability mask for different postures is shown in Figure 4.5. The probabilistic tracking, together with camera calibration and consequent model scaling, provides a scale-invariant, occlusion-free and well-segmented shape of the person as input to the posture classification module.

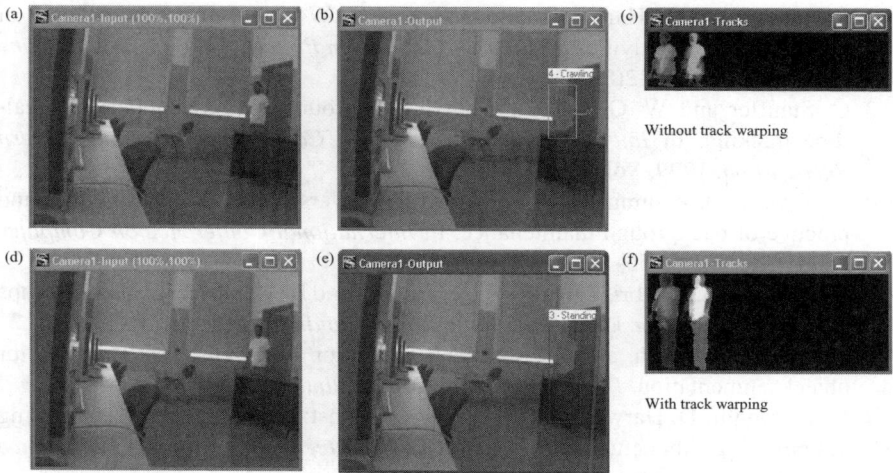

Figure 4.16   *Initial occlusion resolved with track warping: (a,d) input frames; (b,e) output of the posture classification; (c,f) appearance images (left) and probability maps (right)*

Only by taking into account all these aspects is it possible to obtain the very high accuracy performance of the system (in terms of correctly classified postures). We tested the system over more than 2 h of video (provided with ground truths), achieving an average accuracy of 97.23%. The misclassified postures were mainly due to confusion between sitting and crouching postures.

As a test bed for our distributed system, a two-room setup has been created in our lab. An optical sensor is used to trigger the passage of a person between the two rooms. A map of the test environment is shown in Figure 4.15, while the frames extracted from the two cameras in correspondence to a sensor trigger are reported in Figure 4.9. The intersection lines between the floor plane and the entrance plane are automatically drawn by the system exploiting the calibration information. A desk that occludes the legs of a person during his/her entry has been placed in the second room in order to test the efficacy of our approach. In this situation the stand-alone people posture classifier fails, stating that the person is crouching (Figure 4.16 top). Exploiting the described camera handoff module instead, the legs of the person are recovered by the warped appearance mask, and the posture classification produces the correct result (Figure 4.16 bottom).

References

1 R. Cucchiara, C. Grana, M. Piccardi, and A. Prati. Detecting moving objects, ghosts and shadows in video streams. *IEEE Transactions on Pattern Analysis and Machine Intelligence*, 2003;25(10):1337–1342.

2 I. Haritaoglu, D. Harwood, and L.S. Davis. W4: real-time surveillance of people and their activities. *IEEE Transactions on Pattern Analysis and Machine Intelligence*, 2000;22(8):809–830.

3 C. Stauffer and W. Grimson. Adaptive background mixture models for real-time tracking. In *International Conference on Computer Vision and Pattern Recognition*, 1999, Vol. 2, pp. 246–252.

4 K. Toyama, J. Krumm, B. Brumitt, and B. Meyers. Wallflower: principles and practice of background maintenance. In *International Conference on Computer Vision*, 1999, pp. 255–261.

5 S.J. McKenna, S. Jabri, Z. Duric, A. Rosenfeld, and H. Wechsler. Tracking groups of people. *Computer Vision and Image Understanding*, 2000;80(1):42–56.

6 J. Stauder, R. Mech, and J. Ostermann. Detection of moving cast shadows for object segmentation. *IEEE Transactions on Multimedia*, 1999;1(1):65–76.

7 I. Haritaoglu, D. Harwood, and L.S. Davis. Ghost: a human body part labeling system using silhouettes. In *Proceedings of Fourteenth International Conference on Pattern Recognition*, 1998, Vol 1, pp. 77–82.

8 S. Jabri, Z. Duric, H. Wechsler, and A. Rosenfeld. Detection and location of people in video images using adaptive fusion of color and edge information. In *Proceedings of International Conference on Pattern Recognition*, 2000, Vol. 4, pp. 627–630.

9 A. Prati, I. Mikic, M.M. Trivedi, and R. Cucchiara. Detecting moving shadows: algorithms and evaluation. *IEEE Transactions on Pattern Analysis and Machine Intelligence*, 2003;25(7):918–923.

10 C. Cedras and M. Shah. Motion-based recognition: a Survey. *Image and Vision Computing*, 1995;13(2):129–155.

11 D.M. Gavrila. The visual analysis of human movement: a survey. *Computer Vision and Image Understanding*, 1999;73(1):82–98.

12 J.K. Aggarwal and Q. Cai. Human motion analysis: a review. *Computer Vision and Image Understanding*, 1999;73(3):428–440.

13 T.B. Moeslund and E. Granum. A survey of computer vision-based human motion capture. *Computer Vision and Image Understanding*, 2001;81:231–268.

14 W. Hu, T. Tan, L.Wang, and S. Maybank. A survey on visual surveillance of object motion and behaviours. *IEEE Transactions on Systems, Man, and Cybernetics – Part C*, 2004;34(3):334 – 352.

15 L. Wang, W. Hu, and T. Tan. Recent developments in human motion analysis. *Pattern Recognition*, 2003;36(3):585–601.

16 A. Senior. Tracking people with probabilistic appearance models. In *Proceedings of International Workshop on Performance Evaluation of Tracking and Surveillance Systems*, 2002:48–55.

17 R. Cucchiara, C. Grana, G. Tardini, and R. Vezzani. Probabilistic people tracking for occlusion handling. In *Proceedings of IAPR International Conference on Pattern Recognition*, 2004, Vol. 1, pp. 132–135.

18 T. Zhao and R. Nevatia. Tracking multiple humans in complex situations. *IEEE Transactions on Pattern Analysis and Machine Intelligence*, 2004;26(9): 1208–1221.

19 S. Iwasawa, J. Ohya, K. Takahashi, T. Sakaguchi, K. Ebihara, and S. Morishima. Human body postures from trinocular camera images. In *Proceedings of International Conference on Automatic Face and Gesture Recognition*, 2000, pp. 326–331.

20 N. Werghi and Y. Xiao. Recognition of human body posture from a cloud of 3D data points using wavelet transform coefficients. In *Proceedings of International Conference on Automatic Face and Gesture Recognition*, 2002, pp. 70–75.

21 S. Iwasawa, K. Ebihara, J. Ohya, and S. Morishima. Real-time human posture estimation using monocular thermal images. In *Proceedings of International Conference on Automatic Face and Gesture Recognition*, 1998, pp. 492–497.

22 I. Mikic, M. Trivedi, E. Hunter, and P. Cosman. Human body model acquisition and tracking using voxel data. *International Journal of Computer Vision*, 2003;53(3):199–223.

23 E. Herrero-Jaraba, C. Orrite-Urunuela, F. Monzon, and D. Buldain. Video-based human posture recognition. In *Proceedings of the IEEE International Conference on Computational Intelligence for Homeland Security and Personal Safety*, CIHSPS 2004, 21–22 July 2004, pp. 19–22.

24 J. Freer, B. Beggs, H. Fernandez-Canque, F. Chevriet, and A. Goryashko. Automatic recognition of suspicious activity for camera based security systems. In *Proceedings of European Convention on Security and Detection*, 1995, pp. 54–58.

25 H. Fujiyoshi and A. Lipton. Real-time human motion analysis by image skeletonization. In *Fourth IEEE Workshop on Applications of Computer Vision*, 1998.

26 L.R. Rabiner. A tutorial on hidden Markov models and selected applications in speech recognition. *Proceedings of the IEEE*, 1989;77(2):257–286.

27 Q. Cai and J.K. Aggarwal. Tracking human motion in structured environments using a distributed-camera system. *IEEE Transactions on Pattern Analysis and Machine Intelligence*, 1999;21(12):1241–1247.

28 V. Kettnaker and R. Zabih. Bayesian multi-camera surveillance. In *International Conference on Computer Vision and Pattern Recognition*, 1999, Vol. 2, pp. 253–259.

29 L. Lee, R. Romano, and G. Stein. Monitoring activities from multiple video streams: establishing a common coordinate frame. *IEEE Transactions on Pattern Analysis and Machine Intelligence*, 2000;22(8):758–767.

30 S. Khan and M. Shah. Consistent labeling of tracked objects in multiple cameras with overlapping fields of view. *IEEE Transactions on Pattern Analysis and Machine Intelligence*, 2003;25(10):1355–1360.

31 A. Mittal and L. Daviws. Unified multi-camera detection and tracking using region-matching. In *Proceedings of IEEE Workshop on Multi-Object Tracking*, 2001, pp. 3–10.

32 Z. Yue, S.K. Zhou, and R. Chellappa. Robust two-camera tracking using homography. In *Proceedings of IEEE International Conference on Acoustics, Speech, and Signal Processing*, 2004, Vol. 3, pp. 1–4.

33 J. Li, C.S. Chua, and Y.K. Ho. Colour based multiple people tracking. In *Proceedings of IEEE International Conference on Control, Automation, Robotics and Vision*, 2002, Vol. 1, pp. 309–314.

34 J. Krumm, S. Harris, B. Meyers, B. Brumitt, M. Hale, and S. Shafer. Multi-camera multi-person tracking for easy living. In *Proceedings of IEEE International Workshop on Visual Surveillance,* 2000, pp. 3–10.

35 J. Kang, I. Cohen, and G. Medioni. Continuous tracking within and across camera streams. In *Proceedings of IEEE International Conference on Computer Vision and Pattern Recognition*, 2003, Vol. 1, pp. I-267–I-272.

36 S. Chang and T.-H. Gong, Tracking multiple people with a multi-camera system. In *Proceedings of IEEE Workshop on Multi-Object Tracking*, 2001, pp. 19–26.

37 S.L. Dockstader and A.M. Tekalp. Multiple camera tracking of interacting and occluded human motion. *Proceedings of the IEEE*, 2001;89(10):1441–1455.

38 T.H. Chang, S. Gong, and E.J. Ong. Tracking multiple people under occlusion using multiple cameras. In *Proceedings of British Machine Vision Conference*, Vol 2, 2000, pp. 566–576.

39 O. Javed, Z. Rasheed, K. Shafique, and M. Shah. Tracking across multiple cameras with disjoint views. In *Proceedings of IEEE International Conference on Computer Vision,* 2003, Vol. 2, pp. 952–957.

40 A. Mittal and L.S. Davis. A multi-view approach to segmenting and tracking people in a cluttered scene. *International Journal of Computer Vision*, 2003;51(3):189–203.

41 M. Bertini, R. Cucchiara, A. Del Bimbo, and A. Prati. An integrated framework for semantic annotation and transcoding. In Multimedia Tools and Applications, Kluwer Academic Publishers, 2004.

42 R. Cucchiara, C. Grana, and A. Prati. Semantic transcoding of videos by using adaptive quantization. *Journal of Internet Technology*, 2004;5(4):31–39.

43 R. Cucchiara, C. Grana, A. Prati, and R. Vezzani. Enabling PDA video connection for in-house video surveillance. In *Proceedings of First ACM Workshop on Video Surveillance (IWVS)*, 2003, pp. 87–97.

Chapter 5

A general-purpose system for distributed surveillance and communication

X. Desurmont, A. Bastide, J. Czyz, C. Parisot, J-F. Delaigle and B. Macq

5.1 Introduction

The number of security and traffic cameras installed in both private and public areas is increasing. Since human guards can only effectively deal with a limited number of monitors and their performance falls as a result of boredom or fatigue, automatic analysis of the video content is required. Examples of promising applications [1] are monitoring metro stations [2] or detecting highways traffic jams, semantic content retrieval, detection of loitering and ubiquitous surveillance. The requirements for these systems are to be multi-camera, highly reliable and robust and user-friendly. We present a generic, flexible and robust approach for an intelligent distributed real-time video surveillance system. The application detects events of interest in visual scenes, highlights alarms and computes statistics. The three main technical axes are the hardware and system issues, the architecture and middleware and the computer vision functionalities.

The chapter is organised as follows: Section 5.2 describes the requirements of the overall system from the user point of view. Section 5.3 presents a global description of the system and its main hardware characteristics. Section 5.4 illustrates how the hardware and software in a distributed system are integrated via the middleware. Section 5.5 goes deeper into an understanding of the image analysis module, which produces metadata information. Section 5.6 is devoted to the description of a case study, which is that of counting people. Some references to this case study are made all along the document. Finally, Section 5.7 concludes and indicates future work.

5.2 Objectives

As mentioned in Reference 3, a vision system is not just a PC, frame-grabber, camera and software. It is important not to neglect many important issues while finding out requirements. Because the proposed architecture is focused on a computer-oriented architecture, a software life cycle can be applied here. Generic steps for this cycle can be specification/requirements, design, implementation, tests and maintenance. Note that in this chapter only specification, design and an implementation proposal are explained. The specification of the system is an iterative process where the customers give their requirements and the engineering refines these to have clear and exhaustive specifications (e.g. the first requirements of the user, 'count people in the shops and trigger the alarm if there is presence at night' become that the system should be aware in less than 500 ms of the events of people entering or exiting. No more than five miscounts are allowed per day. This would also trigger an alarm in real time in case of over-crowded situations or the presence of people at night). In the scope of video surveillance systems, we can divide the system requirements into four distinct layers:

- Hardware constraints and system issues.
- Software architecture and middleware.
- Computer vision.
- Application functionalities.

The hardware constraints and system issues include the distributed sensors, complexity issues, real time aspects, robustness with MTBF (24 hours a day, 7 days a week), physical protection and safety of equipment, physical condition of use, network, multiple input/output, analogue/digital, existing and new cameras and systems with possible extensions, wired or wireless, ease of integration/deployment, installation, calibration, configuration and running costs like maintenance, easier with the use of standards.

Software architecture and middleware consider modularity, distribution, scalability, standard interfaces and interchange formats, interoperability and portability, reusability, communication, inputs and outputs, co-ordination with other machines and 'plug and play' features and client applications that need to be cross-platform (to be used on smart devices such as enhanced mobile phones or PDAs).

Computer vision aspects involve the robustness of changing context and illumination conditions, handling of complex scenes and situations, generality and ability to migrate to a different system. Because vision algorithms are sometimes very complex, they can have transient computational overload; thus, how they operate in this mode should be investigated.

The application functionalities include the type of functions needed for alarm, storage and content retrieval, qualitative and quantitative performance such as the receiver operating characteristics (ROCs) with false alarm and missed detection, cost effectiveness, image resolution and quality.

Other issues should not be forgotten such as the user interfaces, the skill level needed by an operator, the quality of the documentation and, last but not least, the system having to be cheap enough to be profitable. Note that there are other potential gains from video surveillance like deterrence of theft or violence.

5.3   System overview and description of components

The CCTV system presented in this chapter is based on a digital network architecture and works either with digital or existing analogue cameras. This kind of system can be deployed in a building, for instance, or can be connected to an existing data network. The system is composed of computers or smart cameras connected together through a typical local area network (LAN). The cameras are plugged either into an acquisition board on a PC or directly in the local network hub for IP cameras. A human–computer interface and a storage space are also plugged in this system. The main advantage of such an architecture is its flexibility. The major components of the modular architecture are shown in Figure 5.1. In this implementation, each software module is dedicated to a specific task (e.g. coding, network management). The processing modules are distributed on the PC units according to a configuration set-up. In this set-up, the operator defines the architecture and the scheduling of the distributed system and moreover, they are able to customise the action of the different modules (e.g. the vision processing units require a significant number of parameters). The robustness of the overall system is provided by the logical architecture. It manages the various problems that can arise such as network transmission interruption and a hard drive stopping. Moreover, the overall system performance can be improved by dedicating a co-processor to a vision task.

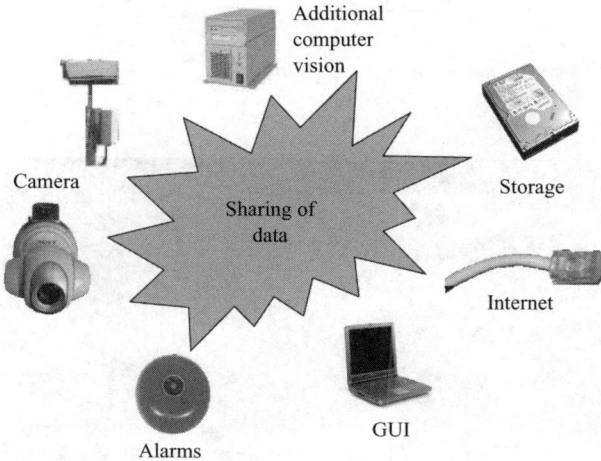

Figure 5.1   The modular architecture

The modules of the software part of the system are explained here. We succesively explain the acquisition module, the codec module, the network module and the storage module. We do not describe here the graphical user interface (GUI) previously reported in Reference 4, but clearly it is an important module from the user's point of view. Section 5.4 deals with the image analysis module.

5.3.1   Hardware

Typical hardware could be smart network cameras (see Figures 5.2 and 5.3) performing acquisition, processing (detection, video compression) and network communication. The equipment is covered with appropriate housing for protection. Usual network components like switches and hubs, storage units or other processing units, a complete PC for user interaction and illumination sub-systems for night vision are also part of the distributed system.

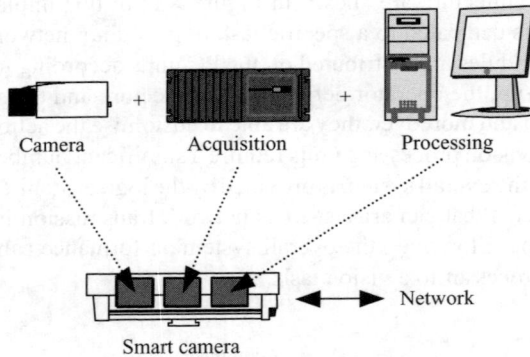

Figure 5.2   The smart network camera principle

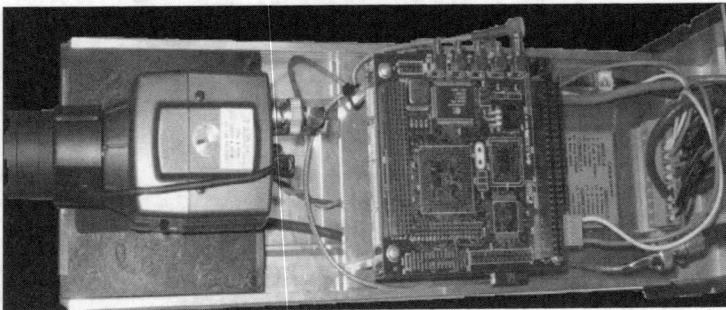

Figure 5.3   The ACIC SA smart camera CMVision (photo by courtesy of ACIC SA)

5.3.2   Acquisition

Acquisition is the generation of a two-dimensional array of integer values representing the brightness and colour characteristics of the actual scene at discrete spatial intervals. The system can currently handle several protocols for image acquisition: IP (JPEG and MJPEG), CameraLink™, IEEE1394 (raw and DV), wireless (analogue, WiFi) and composite (PAL, NTSC, SECAM). A time-stamp is attached to each frame at grabbing time, as this information is useful in the subsequent processing stages. For file replay, we use the Ffmpeg[1] library that handles many codecs.

5.3.3   Codec

The goal of the coding process is to have a good compromise between the compression ratio and bandwidth utilisation. The choice of image quality is done according to the viewer and also to image processing requirements when it occurs after compression. We propose here an MPEG-4 SP compression scheme. Because video surveillance scenes are quite static when cameras are fixed, compression methods suppressing temporal redundancy, such as MPEG-4 SP, are more efficient than classical MJPEG encoding [5]. This technique allows us to transmit up to 20 CIF (352×288) video streams at 25 fps on a typical 100 baseT network.

5.3.4   Network

Having a distributed system implies an efficient use of the available bandwidth. The various modules related to a video signal can operate on several computers. For example, we can have acquisition on computer A, storage on computer B and display on computer C. The basic idea is to use multi-casting techniques to deal with the bandwidth utilisation problem when there are multi-receivers of the same content. This layer can be managed via the middleware (see Section 5.4) through which quality of service (QoS) is ensured.

5.3.5   Storage

The storage module has to deal with the significant amount of data to store, and it must allow 24 hours a day storage. This module has two levels: the first one is a classical storage process using MPEG-4 technology. This level stores a CIF video flow at 25 fps for 3 days on a 40 GB hard drive. We can further improve this number if we allow a two-pass encoder to have a constant quality stream. Up to this point, we have a constant bandwidth stream. Level 1 is an intelligent storage process. It stores only events of interest that the user has defined. The second level saves substantial storage space: typically it could be from 70 to 95 per cent. It also allows fast searching to retrieve stored sequences.

---

[1] http://ffmpeg.sourceforge.net/.

5.4 Software architecture and middleware

5.4.1 Context

Distributed applications need distributed elements. These elements can be processes, data or users. In many cases all these components are heterogeneous and cannot easily communicate together. Before the arrival of client/server architectures, mainframes avoided the issue of a distributed environment by centralising all processes in a single point where terminals were connected. In a client/server architecture, different software processes located on different computers must communicate together to exchange information. In this architecture, clients ask for data or call processes on the server and wait for an expected result. Sometimes clients must communicate with many different servers, but platforms used by these servers are not always the same. In this context the concept of 'inter-operability' appears, which means that a standard way to communicate with a process exists, that is, in a distributed environment elements include a 'built-in' common interface. However, when a distributed environment is built, new problems appear like concurrent accesses (different clients want access to the same data), security (who has access to data or processes) and failure control (what happens when a process fails). Thus, when a distributed system is designed and implemented, these problems must be resolved to ensure the reliability and robustness of the environment. The same mechanism can be used in software running on a stand-alone system (one computer for instance). Typically we can find this in co-operative processes (daemon, etc.), and multi-thread applications. In fact the distributed systems problems and distributed software problems can be addressed in the same way.

To implement these kinds of principles, the software developer must design and implement all the inter-operability layers including the communication layer. As said before, the aim of inter-operability is to provide a 'generic way' to communicate, but if the developer must implement these layers each time, it is a waste of time. In this case it is interesting to have a generic layer that the developer can use and integrate easily to turn back the focus onto the software development itself. The used layer is involved in a framework that all developers (and thus applications) can use always in the same way. Embedded in the application, it hides the system to the software, in other word it is located between the hardware and the software and is named 'middleware'.

5.4.2 Basic architecture and concept

Middleware is placed between the operating system and the software. It is constituted of these parts:

- An API part embedded in the software (the interface between the software and the middleware itself).
- An interface part between the middleware and the platform.
- A runtime part including implementation of API and responsible for the communication between systems.

Table 5.1   The OSI seven layers model

Layer number	Layer name	Description
7	Application	Provides different services to the applications
6	Presentation	Converts the information
5	Session	Handles problems which are not communication issues
4	Transport	Provides end-to-end communication control
3	Network	Routes the information in the network
2	Data link	Provides error control between adjacent nodes
1	Physical	Connects the entity to the transmission media

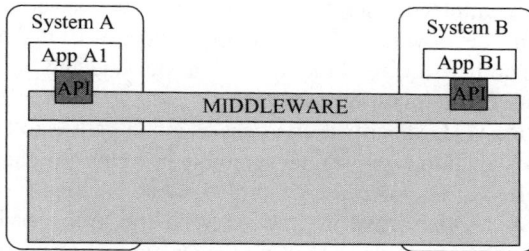

Figure 5.4   System OSI model mapping

The middleware model can be represented as an OSI model (see Table 5.1) in which a level communicates with the same level on a remote system through all sub-layers in a 'transparent' way. In this model, the software sends a request by calling a method on the middleware API, and in the remote software the API calls the related method of the program. Complexity is thus completely hidden (see Figure 5.4).

The API can provide many kinds of functions for communications between each system. The first classical way is the use of remote procedure call (RPC), commonly used in network services and process development. In this configuration the software calls a procedure that is executed on another point (such as other software in the same system or on a remote system). The concept of RPC was ported to remotely call object methods (sometimes called remote method invocation). To allow this call, an object is created in the middleware that maps an object of one (or both) software. The created object is instantiated at each point where the API is used. When it is done, any application can use this object as if it were its own.

Many kinds of middleware implementations exist (like DCE, COM and DCOM), and popular middleware specifications exist also (CORBA, WEB-SERVICE). Because video surveillance systems require communications between modules, the selection of the correct and efficient middleware is a key point of the architecture. To remain in line with the 'standard' requirements of the system, it is preferable to use a specification (and thus an open implementation) of the middleware. To restrict the choice, a comparison is done between CORBA and WEB-SERVICE (other specifications are also interesting but often based on these specifications).

Common object request broker architecture (CORBA) is a standard architecture for distributed systems and was defined by the object management group (OMG). It is based on the client/server architecture, uses a special server called object request broker (ORB) and communicates with the Internet inter ORB protocol (IIOP). The development paradigm is the definition towards the implementation (an interface is written first and then the implementation is done). The CORBA middleware includes some modules or definitions such as interface object repository (IOR, that is a kind of resource locator), interface description language (IDL), some services (naming service, interface repository and trader service). CORBA is a proven technology and has been widely disseminated, but relative to the innovation speed of the information technology domain, it is now an old technology. A lack of CORBA (but which can be resolved) is the ability to extend the network at runtime and outside the network (a known issue with firewall configurations).

The Web Service W3C specification arises as a competitor to CORBA, and is a software system designed to support inter-operable machine-to-machine interactions over a network. It has an interface described in a machine-processable format (specifically WSDL). Other systems interact with the Web service in a manner prescribed by its description using SOAP[2] messages, typically conveyed using HTTP with an XML encapsulation in conjunction with other Web-related standards. Technically, Web Service uses simple object access protocol (SOAP, final version of XML-RPC specification), a protocol over HTTP. The development paradigm is the reverse of CORBA: the implementation is written first and the interface is then generated from the implementation by creating a file written in Web Service description language format (WSDL). Almost the same components of CORBA are defined in Web Service specification, but the main advantage (from our point of view) is the ability to extend easily the network and the possibility to extend the application at runtime (by creating a dynamic object instantiated from the WSDL file). Moreover, Web Service is a young and growing technology and is thus more interesting to investigate.

5.4.3 System and middleware integration

5.4.3.1 Integration overview

As we use middleware, the system is designed following a client/server architecture, where each client is also a server. In this context the concept of 'node' is defined as being hardware connected onto a network. A node can be a camera or a computer dedicated to remote image processing that must be able to communicate with another node. The software is designed to be embedded in the node. In the software architecture, we can regard the image processing runtime as being embedded in a node

---

[2] http://www.w3c.org/TR/soap/.

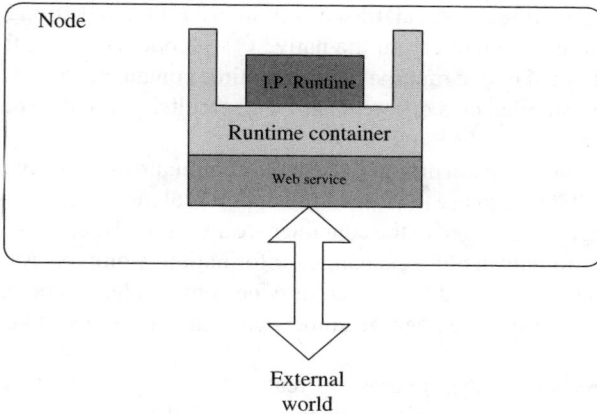

Figure 5.5   Conceptual architecture diagram

and able to be accessed through a middleware. The problem is the way to link the runtime with the middleware. The solution consists in embedding the runtime in a 'runtime container' whose first functionality is to link the runtime and the middleware. We can also add runtime management features to have a better control on the runtime. Ideally we can provide a runtime container that is cross-platform, and in this case it can be pre-installed on a node, which can, as soon as the network is available, wait for the loading and execution of the image processing runtime. At the other side of the runtime container we find the middleware. This module handles messages to and from other nodes on the network. Finally we can provide a representation of the software architecture, as shown in Figure 5.5.

With this structure we can have an independent node that has all the features to enable it to work in a network of inter-communicating nodes. By giving the right signification to the content of the messages exchanged between nodes we can construct image processing runtime that can be aware of information coming from other nodes.

5.4.3.2   Principle of implementation

5.4.3.2.1   Runtime container

To avoid proprietary implementation of the architecture we decided to use the Open Service Gateway initiative (OSGi[3]) specification, which is a 'service oriented' framework. OSGi is a Java-based specification that by itself provides cross-platform features. This framework has already proven to be beneficial on many applications [6]. With OSGi we can provide a service that represents the runtime container, which is based on the Java Native Interface (JNI). The container can call and be called from C++ code in the image processing runtime and offers some management features. By using callback methods, messages exchanged between Java code and native code

---

[3] http://www.osgi.org/.

have to be formatted into a standardised way in order to avoid strong dependencies between the runtime container and the native C++ code. By using this method we can allow any kind of native runtime to run a runtime container. On the other hand, the runtime can be compiled on a different hardware architecture and be directly updated in the container.

Because we have to use a protocol to enable communication between nodes on the network, it is advisable to use the same kind of protocol messages as the middleware, and thus messages exchanged in the container are also SOAP messages. Note also that it can be useful to enhance the exchanged information. Common Alerting Protocol (CAP[4]), a structure designed to be Web Service compatible, can be used to identify nodes, and extra information can be embedded in the messages (like image, sound and resource).

OSGi offers some other interesting features like software life cycle management (such as installation over the network), Jini[5] (a Java-based service that allows functionalities such as discovery, remote invocation, etc.) [7]. By using these functionalities we can obtain a full manageable system, distributed on a network.

In this architecture we have a real-time image processing runtime that is directly designed to be embedded. However, Java is not inherently designed for this purpose (and not for real-time either). To solve this problem, SUN has proposed some specifications, one for embedded software (J2ME[6]) and one for embedded real-time.[7] The advantage of these specifications is that it is possible to execute Java applications on resource-limited hardware or real-time based systems. As OSGi is based on Java, the application that runs on a conventional computer can also run on one of the specific implementations of the Java Virtual Machine (cross-platform).

In summary, we can say that by implementing the runtime container over OSGi and using Web Service features with CAP formatted messages (also used between the container and the native code), we can embed in a node all the software needed to execute image processing and communicate with other nodes in a network for reporting only or to become context-aware [6]. As 'node' is a generic concept, any hardware that can execute this software is a node.

5.4.3.2.2   Specific feature in video surveillance: video streaming

A streaming service can be added to the OSGi framework to allow a node to send video to another node. The streaming is done by a server embedded in the service such as the VideoLan[8] server. When a node needs to send video to another node (for recording for instance), the CAP message sent to the target node contains the resource locator of the video server, and a UDP multicast stream (or unicast) is launched. For video purposes, the quality of service can be obtained by a network supervisor service added

---

[4] http://www.oasis-open.org/specs/index.php#capv1.0.

[5] http://java.sun.com/developer/products/jini/index.jsp.

[6] http://java.sun.com/j2me/index.jsp.

[7] http://jcp.org/aboutJava/communityprocess/final/jsr001/index.html, http://www.rtsj.org/specjavadoc/book_index.html.

[8] http://www.videolan.org.

to the OSGi framework. Note that VideoLan is able to stream over HTTP, MMSH, UDP and RTP, and in this context, the correct selection of required protocol can be done through a management console. A channel information service can be added to the VideoLAN solution based on the SAP/SDP standard. The mini-SAP-server sends announcements about the multicast programs on the network. The network on which the VideoLAN solution is set up can be as small as one Ethernet 10/100 Mb switch or hub, and as big as the entire Internet! The bandwidth needed is 0.5 to 4 Mbit/s for an MPEG-4 stream.

5.4.3.2.3 *System management*

The last problem that we must solve is the way to manage the network itself and collect all events that can appear on the system. The simplest solution is to use the OSGi framework again as a System Server that is responsible for event collections, software life-cycle management, network configuration, etc.

By developing a user interface (which can be embedded in the system server or as a client application of the system server), a user (and also the developer for deployment of the software) can manage the whole system. A good feature of this management interface is the facility to configure the interaction path between nodes (such as to defiine what happens when a person leaves a camera's view range to enter in another camera's one). It is part of the context definition (see Section 5.5.1).

Finally, the system can include a management computer with many cameras with pre-loaded OSGi framework including the runtime container and the middleware. Once a camera is connected to the network, it loads the image processing runtime, installs it in the container and launches its execution. In this system when a new camera is added on the network (for instance using wireless technology), the operator of the system adds the node in the management application, and once the new node (in this case a camera) is detected, the node becomes operational.

5.5 Computer vision

Since the last decade, many algorithms have been proposed that try to solve the problem of scene understanding. The level of understanding varies highly from only detecting moving objects and outputting their bounding boxes (e.g. the Open Source project 'Motion'[9]), to the tracking of objects over multiple cameras, thereby learning common paths and appearance points [8,9] or depth maps and amount of activity in the scene [10].

High-level interpretation of events within the scene requires low-level vision computing of the image and of the moving objects. The generation of a representation for the appearance of objects in the scene is usually needed. In our system, the architecture of the vision part is divided into three main levels of computation that achieve the interpretation (Figure 5.6): image level, blob level and event level.

[9] Open Source project Motion: http://sourceforge.net/projects/motion/.

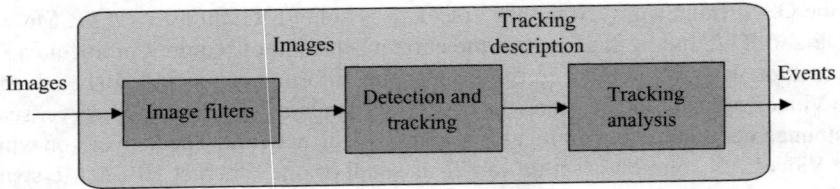

Figure 5.6   Overview of the vision system components

In the general state-space approach, the global problem definition is to predict the state-variables (position of objects, events, etc.) through observations from sensors (cameras, infrared detectors, etc.).

Regarding the tracking algorithm, two different approaches exist: 'bottom-up', for example, where the tracking algorithm tries to match temporally faces located by a detection algorithm, and 'top-down', where tracks are initialised by the detection algorithm and tracking tries to temporally locate certain features of the detected regions of interest (ROIs). We will describe the two approaches.

5.5.1   Context acquisition

In many situations, it is of interest to have calibration of the camera (e.g. for multiple cameras applications). An easy and manual tool for calibration (Figure 5.7) has been developed using the same procedure described in Reference 11. It takes around 5 min to calibrate a camera, and it has to be done every time the camera has moved. We also handle radial deformation via the OpenCV[10] library. Furthermore, an interface allows defining contextual information either globally (information corresponding to many cameras viewing the same scene) or specifically. This contextual information is represented by means of 2D polygons on the image, each of them having a list of attributes: IN_OUT zone, NOISY zone, OCCLUSION zone, AREA_OF_INTEREST zone, etc. This type of information is fed to the image analysis module to help the scenario recognition and the alarms management processes. A 3D context (Figure 5.8) can be set up to describe scenes in 3D (the floor, the walls and the objects present at the beginning of a sequence).

In the case of a multi-camera system (i.e. multi-nodes), it is necessary to give nodes context about other nodes. In this case two modes can be considered: an architect mode (Figure 5.9) where the environment is encoded and a Virtual World (as in Figure 5.8) can be represented for automatic configuration of the network (automatic calibration, tracking correlation between cameras, etc.) and a manual mode (see Figures 5.9 and 5.10) where the user gives manually all the parameters for the configuration of cameras. A deployment module (part of the management application) is able to compute all parameters needed for execution of and installing image processing software (or other software) on each node of the network.

---

[10] http://sourceforge.net/projects/opencvlibrary/.

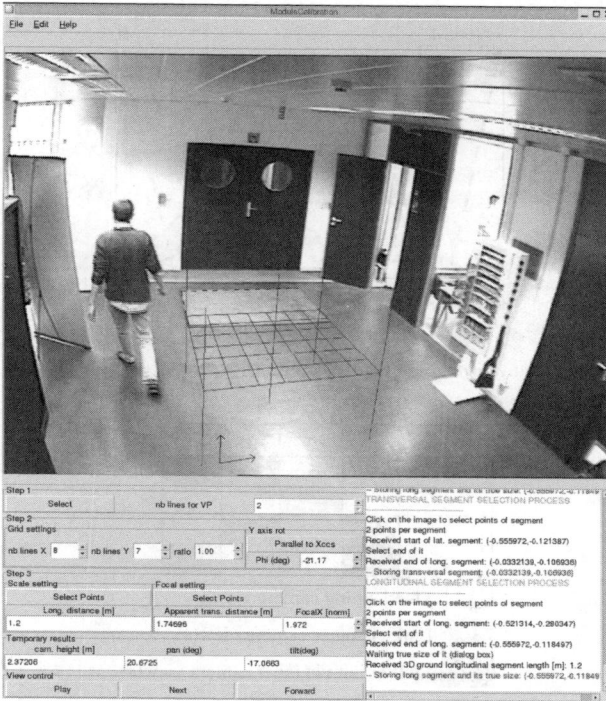

Figure 5.7   Interactive tool for manual calibration

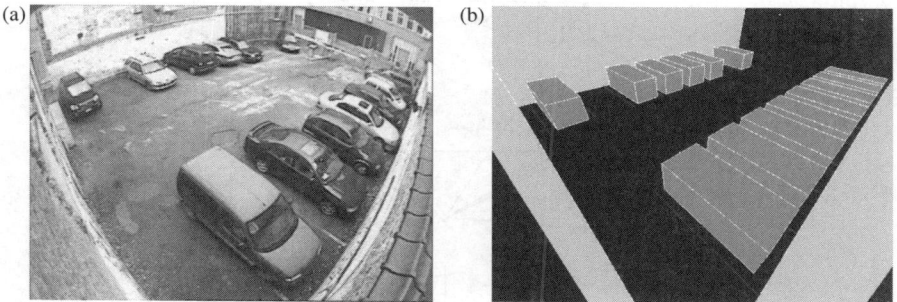

Figure 5.8   (a) Car park: Initial view from a fixed camera. (b) Initial 3D context of the car park

In the case of a moving camera due to a slight vibration (e.g. wind) or pan–tilt–zoom, the image is altered and should undergo a 2D homographic geometric transformation to be integrated back into the global 2D scene view (Figure 5.11). At the current stage of development, the model only handles pure translation (a restricted area of pan–tilt with a small view-angle).

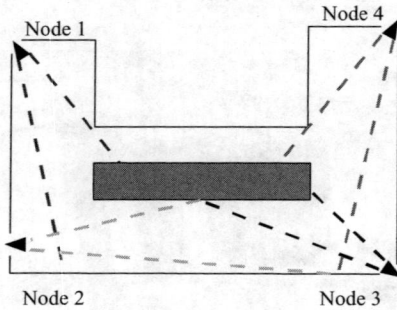

Figure 5.9   Representation of the view range of nodes

Figure 5.10   Representation of links between nodes

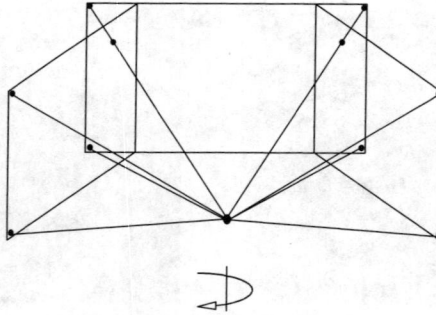

Figure 5.11   Images from left and right views are projected into the middle image plane (scene view)

5.5.2   Image pre-filtering

After acquisition, the current image is filtered in the pixel domain to decrease spatial high-frequency noise. Usually the image is convoluted via a linear filter with a Gaussian kernel, but in our scheme we use an optimised two-pass exponential filter

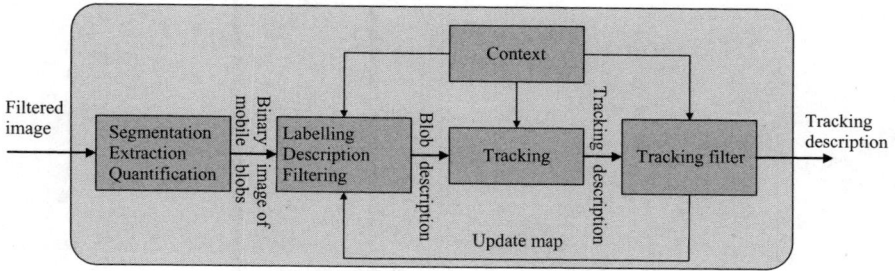

Figure 5.12   Overall processing of the bottom-up tracking

[12] for reduced processing time. The image is then downsised, to reduce the computational cost of the segmentation process. When compressed video streams are used, we do not take into account the loss of quality, as we have not addressed this issue for the moment.

5.5.3   Bottom-up tracking

The overall processing diagram of the bottom-up tracking scheme used is shown in Figure 5.12. The architecture of the bottom-up tracking is divided into two main levels of computation: image level (background evaluation and segmentation) and blob level (description, blobs filtering, matching, tracking description and filtering).

5.5.3.1   Segmentation

Segmentation is the partition of a digital image into multiple regions (sets of pixels), according to some criterion. The common bottom-up approach for segmenting moving objects uses both background estimation and foreground extraction (usually called background subtraction) [13].

The typical problems of this approach are changes of illumination, phantom objects, camera vibrations and other natural effects such as tree branches moving. In the system it is possible to use different reference image models representing the backgrounds (low-pass temporal recursive filter, median filter, unimodal Gaussian, mixture of Gaussians [14,15] and vector quantisation [16]). In the literature, the most basic background model is updated to take small illumination changes into account by

$$B_t = \alpha I_t + (1 - \alpha)B_{t-1}. \tag{5.1}$$

Equation (5.1) is computed for each pixel of the image. α is a parameter of the algorithm, B_{t-1} is the previous background pixel value, I_t is the current image pixel value and B_t is the updated current background value. The foreground is then typically extracted through a threshold T on the difference between B_t and I_t. A pixel is foreground when (5.2) is true:

$$|B_t - I_t| > T. \tag{5.2}$$

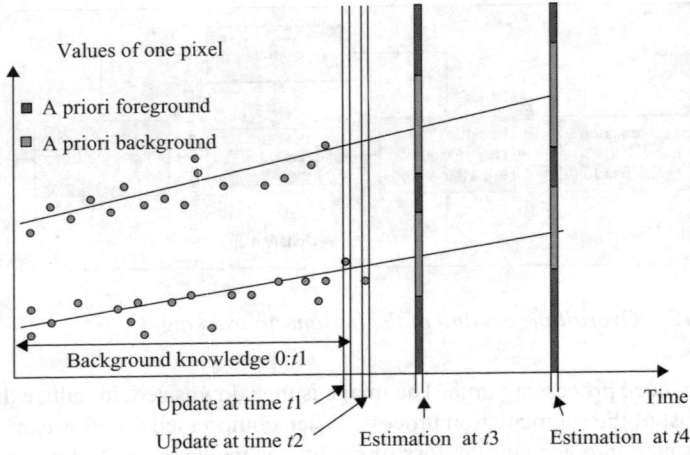

Figure 5.13   Representation of a value history and background model for one pixel in the image

This model is quite simple and can handle small or slow variations of background. However, real applications need more improvement in the model as well as the conceptualisation to be more robust to common noise such as monitor flickering or moving branches in trees. The segmentation is fully configurable as it is possible to choose between different types of background models and parameters on-line (during the running of the system). For all these models, we have the following two methods: estimate foreground and update background. We also add the possibility to predict the background state at a given time and also the possibility of feedback to update selectively the background [17]. Figure 5.13 explains why it is useful to have different stages to update and estimate the background. The dots correspond to values of luminance of the same pixel in a sequence of images for time 0 to $t2$. The estimation for time $t3$ or $t4$ shows the rule of classification of a new pixel between background and foreground (dark=background, bright=foreground). At time $t1$, the model has the history from 0 to $t1$ ($0{:}t1$) and is defined as $B_{0:t1}$. Then there is an update at time $t2$ and it becomes $B_{0:t2}$. It is possible to obtain an estimation of the background at time $t3$ which is $B_{0:t2}(t3)$ and compare it to the value of the pixel at this time I_{t2} to know if the given pixel is to be classified as a foreground or background pixel. As can be seen in Figures 5.13 and 5.15, this approach is, for example, applicable when the background is bi-modal and when the illumination is increasing with time

Let us define $t1 < t2$ two times after image acquisition. The current image I_{t2} has just been acquired at time $t2$ and we have $B_{0:t1}$, the background model updated at time $t1$. The process is divided into several parts (Figures 5.14 and 5.16):

1.   An *a priori* belief map is computed with a normalised distance between I_{t2} and $B_{0:t1}$. This distance is dependent on the background model (e.g. for a mixture of Gaussians it is the difference between the pixel value and the mean of the nearest

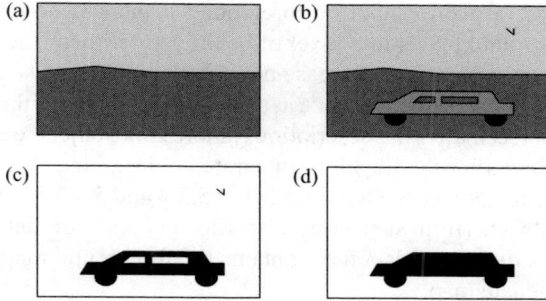

Figure 5.14 (a) Representation of background model at time t1, (b) image from the scene at time t2, (c) foreground at step 2, (d) foreground at step 3 (small blobs and holes removed)

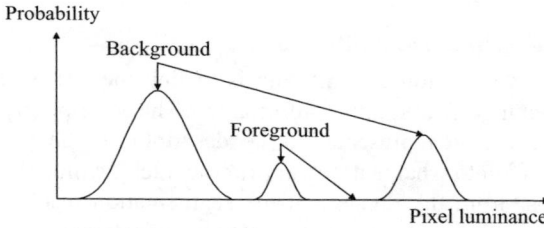

Figure 5.15 Statistical background modelling of a pixel using two Gaussians +1 Gaussian for foreground

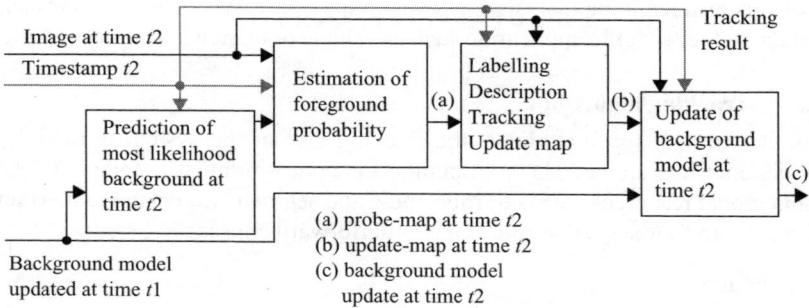

Figure 5.16 Architecture of the segmentation process within the whole architecture

Gaussian). The most probable foreground is then computed: for all pixels in the belief map for which the pixel belief is greater than 0.5, the pixel is *a priori* (without knowledge of the neighbourhood) considered as foreground.

2. If the pixel is inside a closed outline of pixels with a belief greater than 0.5, a neighbourhood rule will consider it as a foreground pixel. A second neighbourhood

rule is applied: a pixel cannot be foreground if there is no path of connected *a priori* foreground pixels to a pixel of belief greater than 0.8. These two rules use a hysteresis approach to decrease noise in the foreground.

3. Then two steps of decision are made at two different stages of the process to filter foreground objects: after the description (ignore zone, object too small, phantom object, etc.) and after the tracking (integrated object, etc.). We do not consider here these processes (see Sections 5.5.3.3, 5.5.3.4 and 5.5.3.5). After them, some foreground objects (from step 2) are discarded or some non-detected objects are added to the structure of data that contains the foreground map. At this time, it is called an update map.

4. The background model is then updated with the image at time *t*2 for regions of the scene that the update map has defined as background.

In Sections 5.5.3.3, 5.5.3.4 and 5.5.3.5, we briefly describe the processes that occur between estimation steps 1 and 4 above.

5.5.3.2   Blobs description and filtering

The aim of blobs description and filtering is to link the processes of foreground extraction and tracking and to simplify information. The description process translates video data into a symbolic representation (i.e. descriptors). The goal is to reduce the amount of information to what is necessary for the tracking module. The description process calculates, from the image and the segmentation results at time t, the k different observed features $f_{k,i}^t$ of a blob i: 2D position in image, 3D position in the scene, bounding box, mean RGB colour, 2D visual surface, inertial axis of blob shape, extreme points of the shape, probability to be a phantom blob, etc. At this point, there is another filtering process to remove small blobs, blobs in an area of the image not considered for analysis, etc. Other model-based vision descriptors can also be integrated for specific application such as vehicle or human 3D model parameters.

5.5.3.3   Tracking algorithm

As is the case for other modules, the tracking part of the system is flexible and fully parametrical online. The configuration is done aiming at a trade-off between computational resources, needs of robustness and segmentation behaviour. Tracking is divided into four steps that follow a straightforward approach:

*   Prediction.
*   Cost matrix.
*   Matching decision(s).
*   Tracks update(s).

Note that there are multiple predictions and cost matrices when the last matching decision is not unique, which is inherent for some matching algorithms in multiple hypothesis tracking (MHT) [18]. Figure 5.17 briefly explains the architecture.

Prediction: The estimation process is very basic. It predicts the blob's features (position, colour, size, etc.). It is integrated as a recursive estimator to handle various

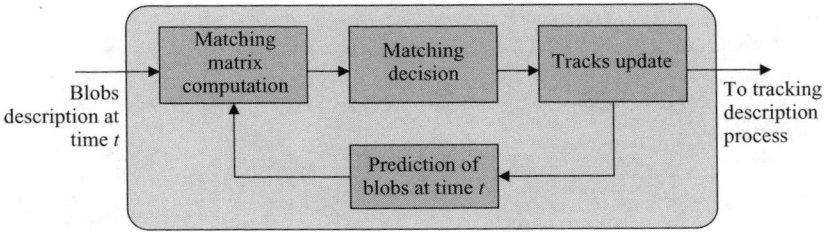

Figure 5.17   Basic tracking architecture

estimator cores like Kalman filter and explicit Euler. The estimation is then taken as the maximum *a posteriori* probability (MAP).

$$\hat{f}_{k,i}^t = \underset{f_{k,i}^t}{\operatorname{argmax}} \, p(f_{k,i}^t | f_{k,i}^{0:t-1})$$

(5.3)

Cost matrix computation: The aim of the matrix is to identify a cost $C_{i,j}$ for a matching between blobs i and j in current and precedent frame (or the estimation of it in current frame). The cost, which is low when the blob looks similar and high otherwise, is performed by a procedure based on all the features of a blob. For example, if images are in colour, the cost depends on colour dissemblance. In fact, $C_{i,j}$ is a linear combination of a weighted distance of each feature (Equation (5.4)). If $\alpha_k = 0$, the distance is not computed. Currently, α_k is defined by an expert during the set-up of the system, but in the future we hope to implement a learning algorithm that maximises some performance aspect of the system.

$$C_{i,j} = \sum_{k=1...N} \alpha_k d_k (F_k^i, F_k^j)$$

(5.4)

Matching: From a given cost matrix, there are many potential hypothesis matrices for the real matching matrix [18]. The library we designed handles three matching algorithms that can be executed (one at a time): 'full multi-hypothesis matching', 'light multi-hypothesis matching' and 'nearest neighbour matching':

- Full multi-hypothesis matching: all hypotheses of matching are tested, the K best global hypotheses are kept. It takes a significant amount of computation time. For example, from the K best global previous hypotheses in a case of n blobs per frame, the number of hypotheses to test is $K \times 2^{n \times n}$. In practice, some hypotheses are more likely to occur than others, and hence we define the two heuristics below.
- Light multi-hypothesis matching: it reduces the number of matching hypothesis matrices. For a given blob in the current image, it only performs matching hypotheses between the N blobs in the precedent image that have the lowest costs.
- Nearest neighbour matching: this should be the simplest matching algorithm. Every blob from the precedent frame is matched to its nearest blob (lowest cost) in the current frame. Furthermore, every blob from the current frame is matched

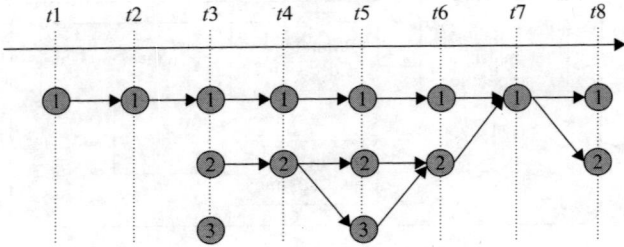

Figure 5.18   *Internal tracking result description. t1–t8 are the time-stamps. Circles show objects in the image at time t and arrows show matchings between objects in different frames*

to its nearest blob in the precedent frame. However, if the cost between two blobs exceeds a threshold given as a parameter, the match is removed.

Note that at each iteration the matching algorithm can be permuted with another. Note also that an ambiguity multi-disjoint-hypothesis tracking can be developed to efficiently reduce the combinatorial dependence [18].

5.5.3.4   Tracking description and filtering

The aim of tracking description and filtering is to provide an interface between the tracking and the analysis processes and to simplify information. The tracking description converts the internal tracking result to a graph (Figure 5.18), and then it adds some useful information to the matching data. It computes the time of life of every blob of a track (i.e. the duration of the track from its first occurrence to the specified blob), the time before 'death' (i.e. the duration of the track to disappearance of the specified blob). It also describes a piece of track restricted to a small area as is the case for a stationary object.

The grammar of the tracking description of blobs behaviour includes apparition (new target), split, merge, disappearance, stopped, unattended object, entering a zone and exiting a zone. For apparition, there is no 'confirmTarget' as in Reference 19 because the filter 'smalltrack' removes tracks that last for less than a fixed duration.

Tracking filtering is performed on the tracking description output. It is as necessary as the other filters of the vision system. As usual the filter is used to remove noise. At this level of processing, it can exploit temporal consistency. We describe below some types of filters that can be used in cascade. Because the tracking description is a construction built piece by piece during the progression of the video sequence it can process online or off-line.

smalltrack detects and removes tracks that last for less than a fixed duration. The duration of the track is the delay between apparition and disappearance. A similar filter is used in Reference 20. This kind of noise comes after the segmentation has (incorrectly) detected noise in the image as a blob. Figure 5.19 shows how the object labelled 3 at *t*3 has been erased by this filter.

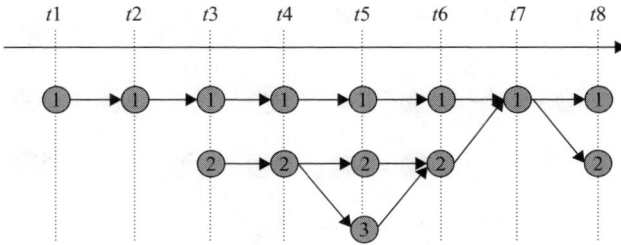

Figure 5.19   Output of smalltrack filter

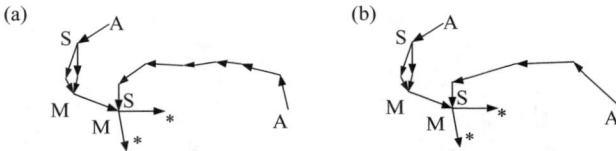

Figure 5.20   (a) Raw tracking description; (b) tracking description filtered by simplifycurvetrack. Symbols A, D, S, M and * mean, respectively, apparition, disappearance, split, merge and 'object in current frame'

simplifycurvetrack simplifies tracks by removing samples of blobs that give poor information (i.e. if we delete it, we can interpolate it from other parts of the track). It is done by removing a blob instance if it is near the interpolated track made without taking it into account. This could be regarded as a dynamic re-sampling algorithm. Figure 5.20 illustrates graphically the difference with and without this filter. Figure 5.21 shows the output in tracking description.

simplifysplitmerge removes one part of a double track stemming from a split and then a merge. This kind of noise comes when the segmentation detects two blobs when in all likelihood there is a unique object. It is done by finding when two blobs result from a split and will merge again in the next frame. This filter also splits tracks that have been merged because of proximity of objects in the image. To do so it matches objects before and after the merge/split with features similarities. Figures 5.22 and 5.23 show typical results.

Another filter removes tracks that start or end in a defined region of the image. Another one removes the tracks of phantom objects. Unfortunately, giving a full description of these filters is outside the scope of this chapter.

5.5.3.5   Feedback to image level

At this stage it is possible to make high-level decisions that would be useful for the segmentation stage. For example it is possible with a heuristic to find so called phantom objects and tell the segmentation to integrate the corresponding blob. Typically the segmentation output is a map of blobs. A blob could represent a real object or a phantom (an apparent object that is the result of incorrect segmentation). One simple

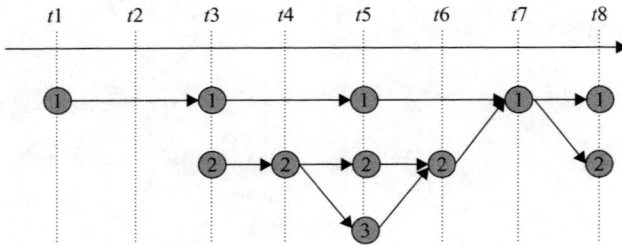

Figure 5.21   *After filter simplifycurvetrack a track has been simplified by removing some object instances*

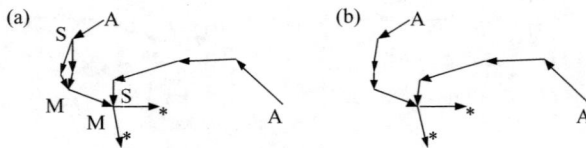

Figure 5.22   *(a) Raw tracking description; (b) tracking description filtered by simplifysplitmerge*

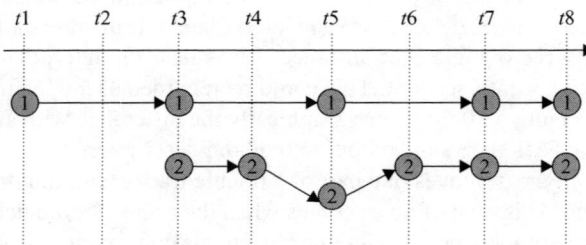

Figure 5.23   *Output tracking after simplifysplitmerge*

heuristic can detect around 90 per cent of these blobs and it uses the edge mean square gradient (EMSG). When this measure is below a certain threshold, the blob is considered as phantom. An example is given in Figure 5.24.

Definition of EMSG

We define the edge of a blob as the set of pixels of the blob which have at least one neighbour not included in the blob (in connectivity 4). The length of the edge is defined as the number of pixels of the edge. We define the inner edge and the outer edge as the pixels next to edge pixels inside and outside the blob, respectively. For each pixel at the edge, we find a pixel at the inner edge and a pixel at the outer edge. The difference in value of these inner and outer pixels is the projection of the gradient of value along and orthogonal to the axis of the local edge front. We assume

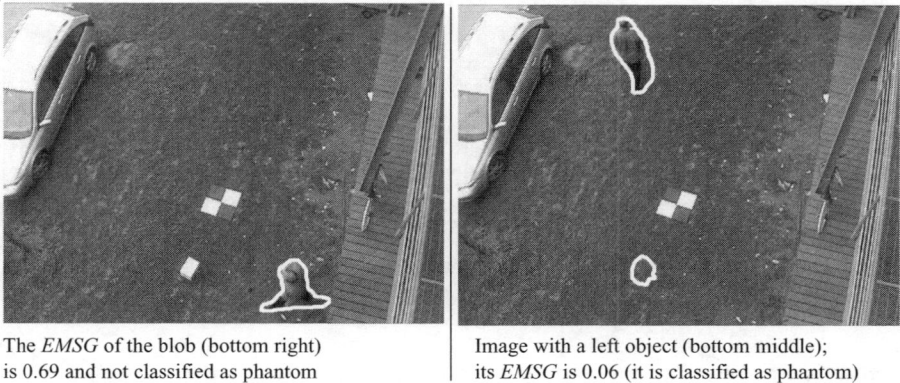

| The *EMSG* of the blob (bottom right) is 0.69 and not classified as phantom | Image with a left object (bottom middle); its *EMSG* is 0.06 (it is classified as phantom) |

Figure 5.24 Representation of the EMSG of a phantom blob

that in the background the gradient is typically low, but orthogonal to the edge of an object it is high, except if the object looks locally like the background. In this scope we use RGB images and define the colour distance of two pixels P and P' as SDISTANCE (Equation (5.5)).

$$\text{SDISTANCE } (P, P') = (P'r - Pr)^2 + (P'g - Pg)^2 + (P'b - Pb)^2 \qquad (5.5)$$

We define the EMSG (Equation (5.6)) as the quadratic sum of the distance of the inner and outer pixels of the edge divided by the length and the range of values.

$$\text{EMSG} = \frac{\sum_{p \in \text{edge}} \text{SDISTANCE}(P_{\text{inner}}, P_{\text{outer}})}{\text{length} \cdot 3 \text{ range}^2}. \qquad (5.6)$$

5.5.4 Top-down tracking

In this section we describe top-down tracking. It consists of estimating the position and shape of an object of interest (the target) in the current video frame, given its position and shape in the previous frame. It is thus by definition a recursive method, see Figure 5.25.

The most restrictive hypothesis of the approach described in Section 5.5.3 is the fixed background requirements. Although moving background estimation is possible, it is usually very time consuming and leads to noisy results. An alternative is to use a top-down approach. In contrast to the bottom-up approach, in the top-down approach, hypotheses of object presence are generated and verified using the data from the current image. In practice, a model of the object is available and is fitted to the image. The parameters of the transformation are used to determine the current position and shape of the object.

A few techniques representative of the top-down approach for tracking are described below. They differ in the type of features extracted from the frame as well as in the tracking technique (deterministic or stochastic). Since none of the methods

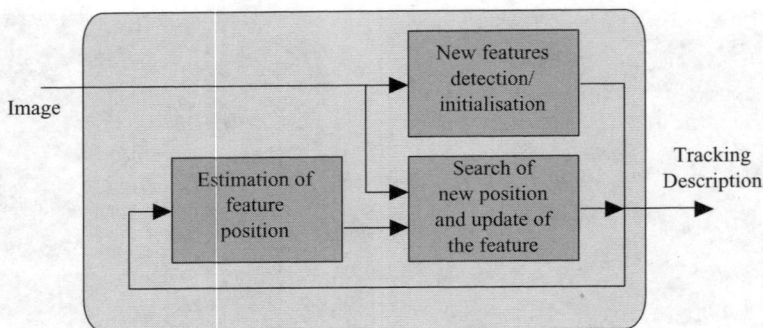

Figure 5.25   The iterative flow chart of top-down tracking

works perfectly in all cases, the choice for one particular method depends on the application requirements.

In all cases a model of the target is assumed to be available. A human operator can provide the model manually: the kind of application is 'click and track'. The operator chooses a person or object to track in this image. This image region is then used to compute the target model. Alternatively, it can come from a detection step: once a blob is detected using the bottom-up techniques described in Section 5.5.3, a model or a template of the target can be extracted automatically from that blob.

Many different target model types have been proposed. Rigid templates are useful if large deformations are not expected. Colour histograms may be more appropriate in the case of non-rigid motion. However colour also has serious drawbacks:

- It varies drastically with illumination.
- Colour is not available when using infrared cameras.
- Colour is a poor feature when people wear 'dark' clothes. Contours have been proposed as well.

The appearance of objects in images depends on many factors such as illumination conditions, shadows, poses and contrast effects. For this reason, it is often necessary to use adaptive models, that is, models of targets that can evolve in order to adapt to appearance changes. If the target model is parametrised as a vector v_t, the update of the target model is done using Equation (5.7), where v_{obs} is the target parametrisation as observed in the current frame. Note the similarity with the background adaptation equation (Equation (5.1)).

$$v_t = (1 - \alpha)v_{t-1} + \alpha v_{obs}. \tag{5.7}$$

In what follows, three tracking methods are presented: the Lucas–Kanade algorithm [21,22], the mean-shift algorithm [23] and the particle filter tracking [24,25]. The Lucas–Kanade and the mean-shift algorithms are both tracking methods formulated as a deterministic optimisation problem. They differ essentially by the features forming the target model. The particle filter based tracking, in contrast, is a stochastic search.

5.5.4.1   Template-based tracking: the Lucas–Kanade algorithm

Suppose the target can be modelled as a region of pixels denoted by $T(x,y)$ called the template. This region can be defined manually by a human operator or from a previous detection step. Suppose also that the position of the target in the previous frame is known. The Lucas–Kanade tracking algorithm is based on the hypothesis that the new position of the target minimises the difference between the current frame and a deformed version of the template. This is often called the brightness constraint. Let f_p be the transformation on the template at frame $t - 1$ (e.g. involving scaling or affine transformation) that matches the template to the current appearance of the target. The transformation is parametrised by a vector p. The brightness constraint can be written as in Equation (5.8), where E is an energy term to be minimised with respect to δp and W is a pixel window. The new position and shape of the object at time t is then $p+\delta p$. In order to minimise E, an iterative gradient descent is performed.

$$E(\delta p) = \sum_{x,y \in W} (I(f_{p+\delta p}(x,y)) - T(x,y))^2. \tag{5.8}$$

The LK algorithm can be regarded as a search for the image patch that resembles most of the template. The search is formulated as a deterministic optimisation where the difference between a template and the current frame is minimised.

Note that instead of starting the gradient search from the previous target position, the dynamics of the target can be modelled and the search started from a predicted position, using, for example, a Kalman filter. It can be shown that the energy function E that is minimised during the gradient descent is not convex. This means that the search procedure may converge to a local minimum of the function. If this happens, the algorithm is likely to lose the target track. Moreover, if the target is occluded, even partially, the algorithm may converge to an incorrect position and lose track.

5.5.4.2   Colour tracking: the mean-shift algorithm

If large deformations of the target are present, the LK algorithm will likely fail (since only small deformations of the template are modelled). To cope with large deformations, the target can be modelled using its colour histogram instead of a pixel grid-based template. The colour histogram of an object is unaffected by rigid and non-rigid deformation. However colour has drawbacks already mentioned above. The computation of the histogram is done in an ellipse centred on the target. To increase robustness against partial occlusions, each pixel colour is weighted by its distance to the ellipse centre in such a way that pixels on the boundary have less influence on the histogram than pixels close to the target centre.

A measure of similarity between colour histograms known as the Battacharya coefficient is used to match the target model with the current observation. The Battacharya coefficient between histogram p_u and q_u is expressed as in Equation (5.9), where u is the colour index. Again, tracking amounts to maximising ρ with respect to position (x, y) in order to determine the position of the target in the current frame.

$$\rho(x, y) = \sum_u \sqrt{p_u q_u}.$$

(5.9)

Instead of computing an iterative gradient ascent, the optimisation is based on mean-shift iterations. It is known that the mean of a set of samples drawn from density points is biased towards the nearest mode of this density. This bias is called the mean-shift and it points in the direction of the gradient. Thus it gives a simple and efficient way of estimating the gradient of ρ. By moving the target candidate position along the mean-shift vector and iterating until the mean-shift displacement is negligible, we perform a gradient ascent of the Battacharya similarity measure. The mean-shift iterations allow the ellipse to be found in the current frame, which has the colour distribution closest to the target colour distribution in the sense of the Battacharya measure. Again, the search procedure may converge to a local maximum of the similarity function, in which case the algorithm is likely to lose the target track.

5.5.4.3   Particle filter tracking

The two previous methods perform a deterministic search in the current frame for the best target candidate in the sense of a similarity measure. Clearly, only the best hypothesis of presence of the target is kept. However, the appearance of the target may be different from frame to frame. In the worst case, the target may be completely occluded for a certain time. In that case, deterministic approaches fail since only one hypothesis is considered. Particle filter (PF) tracking attempts to remedy this problem: a target presence probability in the image is modelled as a finite set of weighted samples or particles. A given particle contains a target position hypothesis as well as other parameters useful for modelling the target such as the size of the target, its shape, intensity and other kinematics parameters. The weight is simply a scalar representing the importance of a given particle. One iteration of the PF can be intuitively understood as follows. From the previous iteration, a set of weighted particles is available. Depending on the dynamic process used for modelling target motion, each particle is randomly perturbed in accordance with the dynamic model. Then each particle receives a weight depending on how well the target model matches the current frame. Estimates for the target parameters (position, size, etc.) can be computed simply by computing the sample mean of all the particles and taking into account their weight.

In order to keep diversity in the particle weights, it is necessary to re-sample the particles. In this procedure new particles with uniform weights are drawn (with replacement) from the particle set, with the constraint that particles with large weight have a greater probability for being drawn. Particles with large weights generate many 'children', and particles with too small a weight may not be drawn at all.

It turns out that the particle filter approach implements the Bayesian recursive estimation. It can be seen as a generalisation of the Kalman filter in the case of non-linear and/or non-Gaussian processes [24].

These two steps can be seen as a stochastic search for the target in the current frame, as particles are randomly spread in the image. The second fundamental feature of the PF is its ability to consider many hypotheses at the same time, since particles with small weight are kept in the next iterations (those particles will eventually disappear if the hypothesis is not verified, but they tend to survive in case of brief total occlusion, allowing maintenance of the track of the object).

Formally, the particle filter requires the definition of two probability densities: the proposal density and the observation density. For tracking applications, the proposal density is usually the transition density $p(x_t|x_{t-1})$, where x_t is the state vector at time t, that is, the probability of transition from a given state to another. As for the observation density $p(z_t|x_t)$, it relates the current observation to the target state and must be provided by the designer. The particle filter is then implemented as follows:

1. Prediction step: using the previous set of particles $\{x_{t-1}^i\}$, draw N particles from the transition density $x_t^i \rightarrow p(x_t|x_{t-1}^i)$.
2. Update step: weight the particle using the observation density $w_t^i = p(z_t|x_t^i)$.
3. Normalise the weights so that $\sum_i w_t^i = 1$.
4. Compute state estimate $\bar{x}_t = \sum_i w_t^i x_t^i$.
5. Re-sample step: re-sample the particle in order to have uniform weights.

Particle filters were originally introduced in vision problems for contour tracking [25]. Since then they have been extended to other features. For example, in Reference 26 a colour based particle filter is introduced. It can be considered as a probabilistic version of the mean-shift tracking. Moreover, Reference 27 proposed a template-based tracking method using particle filters. In this case the particles contain the transformation parameters to match the target template with the current observation. Again, this can be seen as a probabilistic version of the Lucas–Kanade approach.

5.5.5 Tracking analysis and event generation

The tracking analysis is a process that receives the tracking description. It can find pre-defined patterns like objects entering from a defined zone of the image and exiting by another one, or objects which have exceeded a certain limit speed, or also objects that have been stationary for a minimum time which stem from another moving object. Figure 5.26 shows this particular pattern of tracking description. In our application we use this last pattern to recognise that 'someone is taking an object'. The processing of this module looks into the tracking description graph (e.g. Figure 5.23) to find pre-defined event patterns. A similar approach has been described in Reference 19.

5.5.6 Metadata information

As mentioned by Nevatia *et al.* [28], an ontology of events requires means of describing the structure and function of events. The structure tells us how an event is composed of lower-level states and events. This structure can involve a single agent or multiple agents and objects. Properties, attributes and relations can be thought of as states. An event is a change of state in an object. Events generally have a time

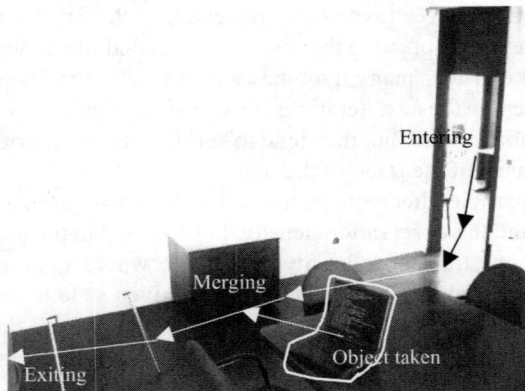

Figure 5.26   Tracking description pattern for 'stolen object'

<?xml version="1.0" encoding="UTF-8"?> <Description> <time>2003/08/08@17:34:29.680000</time> <PositionTagX>X135</PositionTagX> <PositionTagY>Y45</PositionTagY> <ExtentTagX>X13</ExtentTagX> <ExtentTagY>Y32</ExtentTagY> </Description>	<?xml version="1.0" encoding="UTF-8"?> <event_history> <event> <time>2003/08/08@17:34:16.920000</time> <name>unattended_object</name> <description></description> <parameters/> </event> </event_history>

Figure 5.27   Left: XML representation of a segmented object described with the bounding box. Right: XML of scenario detection

instant or interval when or during which they occur. Events generally have locations as well, inherited from the locations of their participants.

The output of the image processing is shared with other cameras through the middleware. The basic structure of this metadata is XML. It describes visually what happens within the image (left-hand side of Figure 5.27) and could also integrate high-level scenario representation (right-hand side of Figure 5.27).

5.5.7   Summary

The global approach is summarised in Figure 5.28, where the three processing axes are shown: description, filtering and intelligence. The first axis includes the acquisition of the image (as in Section 5.3.2) and provides segmented images (Section 5.5.3.2), tracking results (Section 5.5.3.4) and events (Section 5.5.6). The second axis is the filtering of the content (see Sections 5.5.2, 5.5.3.2 and 5.5.3.4). The third axis is

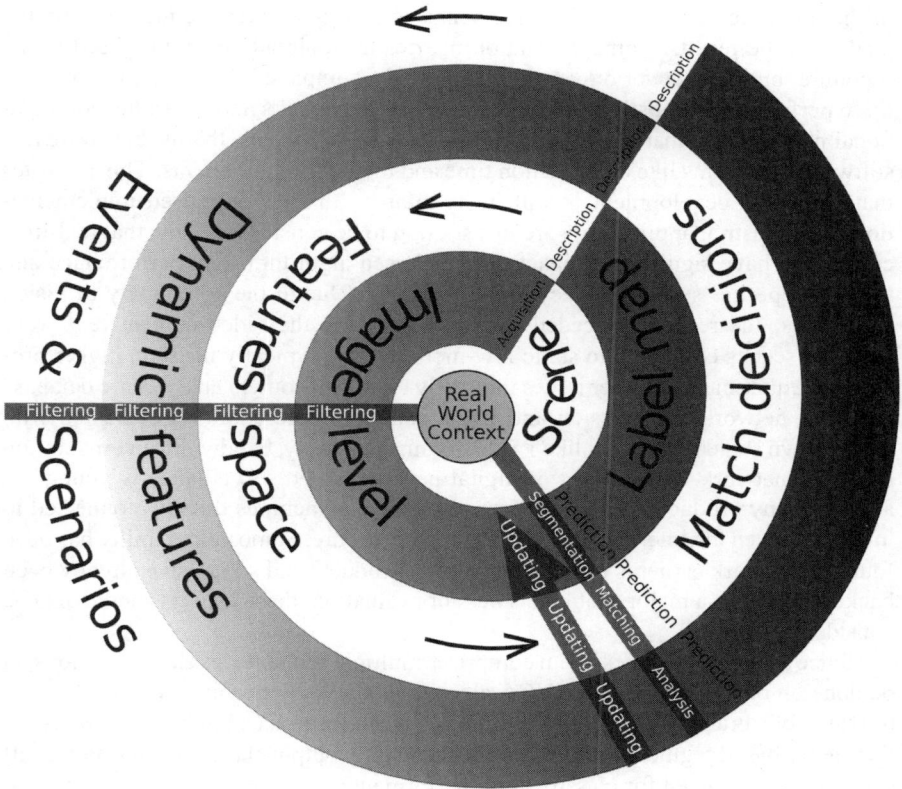

Figure 5.28   Global view of computer vision and analysis

the intelligence (see Sections 5.5.3.1, 5.5.3.3, 5.5.4 and 5.5.5), which is usually more complex than the two other axes. This intelligence axis can usually be divided into two steps: first, the prediction of model behaviour and secondly the update in accordance with observation. Note that the update task could receive feedback (e.g. as in Section 5.5.3.5) from other processes or even a human (see the arrow in the figure). Note also that for top-down tracking, segmentation and tracking are merged into a single process. The two axes, cognition and filtering, use directly the context of the scene that is represented in the middle of the spiral as 'real world context' (see Section 5.5.1).

5.6   Results and performance

As reminded in Reference 29, apart from functional testing, there are several other reasons for evaluating the video content analysis (VCA) systems: scientific interest, measuring the improvement during development, benchmarking with competitors, commercial purposes and finally legal and regulatory requirements. However, most of the literature describing VCA algorithms cannot give objective measures on the quality of the results. Some evaluation metrics have already been proposed in the literature, but most cover only a limited part of a complete VCA system. However, these performances mostly refer to the quality of the results and not to the computational performance that is very relevant for real-time systems, that is, hardware and software constraints like computation time and memory requirements. This indicates that algorithm development is still in its infancy, that is, optimised implementations for industrial applicability are just starting to be considered. Note that real-time constraints have significant impact on the chosen algorithm with performance and timing properties such as latency and throughput. Due to the complexity of vision algorithms, the resources needed usually depend on the video content (e.g. very dynamic scenes compared to static low-motion video typically result in higher processing requirements). Other issues indirectly related to content analysis are database retrieval, network transmission and video coding. These could be evaluated with well-known standard metrics like PSNR for image quality, bandwidth and maximum delay for network. The market of digital network video surveillance is young but already many products exist, but because they are sometimes directly connected to intranet or even the Internet, they need to be very secure. Some vulnerability has been found in network cameras or video servers in products and some have already been hacked, so this is a major problem. Thus for evaluation, these aspects should also be considered.

Since a complete VCA system comprises multiple semantic levels, evaluation can be done on one or more levels. VCA algorithms that only perform segmentation of moving objects and no tracking over multiple video frames, can only be evaluated on their pixel-based segmentation. Algorithms that only output alarms (e.g. car detected) can only be evaluated for classification performance, but not for their segmentation quality. Therefore, first there needs to be defined what exactly should be evaluated for each semantic level. For each level, different metrics are required (there is a good review in Reference 29). Most evaluation proposals are based on pixel-based segmentation metrics only [30–38]. Oberti *et al.* [39] also consider object level evaluation and Mariano *et al.* [40] consider object tracking over time. Nascimento and Marques [41] consider object-based evaluation and introduce splitting and merging of multiple objects. Evaluation of object trajectories is only proposed by the authors of References 42–44. Most proposals require manually annotated ground truth, while a limited set of proposals apply performance evaluation using metrics that do not require ground truth [37,45,46].

During evaluation, values are calculated for each metric, for a certain time interval. Combining these results over the total time interval (of one video sequence or multiple sequences) can be applied in various ways. Some authors show histograms of results over time, while others summarise measures over time. Other metrics require statistical analysis methods like the mean or median. With low variance, the mean gives typical working performance. The maximum error rate in any of the tested cases can also be of interest to prove the limits of the system. From these results also other measurements relevant to industrial deployment can be computed, such as the mean time before failure (MTBF).

5.7 Application case study

To illustrate the interest of the proposed system architecture in real case studies, we describe some scenarios of deployment for indoors or outdoors applications.

The case study has already been introduced. The hardware is CMVision from ACIC S.A. (see Figure 5.3); the middleware was described in Section 5.4. The vision part of the high-level processing modules (detection of activity, abandoned object, missing object, people running too fast, car park management, etc.) can be run as options in accordance with the end-user application. Here we decide to focus on one case: people tracking and counting in indoor scenes. This module should be able to measure the number of people moving from a region of a scene to another. The set-up of this module requires the definition of the context, (i.e. a region of interest), where tracking needs to work on the counting zone, where counting is achieved in accordance with given directions. We can also specify authorised or non-authorised displacements (e.g. to raise an alarm when somebody tries to exit through an entrance door). Figure 5.29 shows a possible camera position (in this case we place three cameras). Figure 5.30 shows what the front camera sees.

Here is an example of concrete deployment of such a technology:

- Deploy the power and network wiring in the building.
- Install the set of smart network cameras connected to the network.
- The cameras automatically learn the topology of the network (i.e. there are three cameras).
- Connect a client to this network (let us say a notebook with a standard operating system installed (MacOS, Windows or Linux).
- Define the service needed (here this is counting and tracking).
- Upload from a CDROM or the Internet the service functions needed with correct licences.
- Define the topology of the building and location of cameras.
- Start the system and unplug the user client.

Note that the definition of the topology and requirements is usually made by an expert, and it could be done from a remote location. Future versions of functions

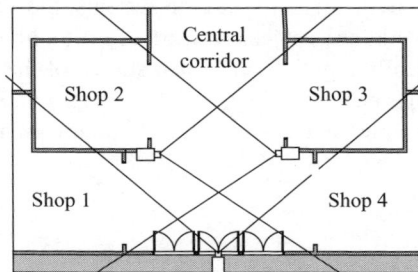

Figure 5.29   The layout of the shopping centre

Figure 5.30   View from the front camera

could automatically be downloaded from the Internet if needed after the installation. Failure detection could also directly call a person to change the faulty component in the distributed system. Cameras could be aware of the context by receiving information from others cameras connected to the same network.

5.8   Conclusions

In this chapter, we have proposed an approach for a distributed video surveillance platform that can provide the flexibility and efficiency required in industrial applications. The scope of issues is divided into three main parts: hardware, middleware and computer vision. We propose to use new standard middleware to overcome the complexity of integrating and managing the distributed video surveillance system. We described parts of the image analysis modules focusing on segmentation with background differencing but also on tracking and analysis. We then showed performance evaluation techniques.

Future work will be in the main axis of the systems: the aim to inter-connect the video surveillance network to the safety network of the building to detect smoke and fire or also to automatically highlight the video stream of the neighbourhood of the sensor where an alarm has been triggered. We also plan to continue investigating new vision modules, for example, better segmentation and tracking methods.

Acknowledgements

This work has been supported by the Walloon Region (Belgium). We also thank Bruno Lienard for his helpful inputs.

References

1 A. Cavallaro, D. Douxchamps, T. Ebrahimi, and B. Macq. Segmenting moving objects : the MODEST video object kernel. In *Proceedings of Workshop on Image Analysis for Multimedia Interactive Services (WIAMIS-2001)*, 16–17 May 2001.

2 F. Cupillard, F. Brémond, and M. Thonnat. Tracking groups of people for video surveillance. In *Proceedings of the 2nd European Workshop on Advanced Video-Based Surveillance Systems*, Kingston University, London, September 2001.

3 B.G. Batchelor and P.F. Whelan. *Intelligent Vision Systems for Industry*. University of Wales, Cardiff, 2002.

4 B. Georis, X. Desurmont, D. Demaret, S. Redureau, J.F. Delaigle, and B. Macq. IP-distributed computer-aided video-surveillance system. In *First Symposium on Intelligent Distributed Surveillance Systems*, IEE, London, 26 February 2003, pp. 18/1–18/5.

5 P. Nunes and F.M.B. Pereira. Scene level rate control algorithm for MPEG-4 video coding. In *Visual Communications and Image Processing 2001*, San Jose, *Proceedings of SPIE*, Vol. 4310, 2001, pp. 194–205.

6 T. Gu, H.K. Pung, and D.Q. Zhang. Toward an OSGi-based infrastructure for context-aware applications. *IEEE Pervasive Computing*, October 2004: 66–74.

7 T. Nieva, A. Fabri, and A. Benammour. Jini technology applied to railway systems. In *IEEE International Symposium on Distributed Objects and Applications*, September 2000, p. 251.

8 D. Makris, T.J. Ellis, and J. Black. Learning scene semantics. In *ECOVISION 2004, Early Cognitive Vision Workshop*, Isle of Skye, Scotland, UK, May 2004.

9 T.J. Ellis, D. Makris, and J. Black. Learning a multi-camera topology. In *Joint IEEE International Workshop on Visual Surveillance and Performance Evaluation of Tracking and Surveillance (VS-PETS)*, ICCV 2002, Nice, France, 2003, pp. 165–171.

10 D. Greenhill, J. Renno, J. Orwell, and G.A. Jones. Learning the semantic landscape: embedding scene knowledge in object tracking. *Real-Time Imaging*, January 2004, Special Issue on Video Object Processing for Surveillance Applications.

11 A. D. Worrall, G. D. Sullivan, and K. D. Baker. A simple, intuitive camera calibration tool for natural images. In *Proceedings of 5th British Machine Vision Conference*, University of York, York, 13–16 September 1994, pp. 781–790.

12 J. Shen and S. Castan. An optimal linear operator for step edge detection. *CVGIP*, 1992;54:112–133.

13 A. Elgammal, R. Duraiswami, D. Harwood, and L.S. Davis. Background and foreground modeling using nonparametric kernel density estimation. *Proceedings of the IEEE*, 2002;90(7):1151–1163.

14 C. Stauffer and W.E.L. Grimson. Adaptive background mixture models for real-time tracking. In *Proceedings of IEEE Conference on Computer Vision and Pattern Recognition*, Fort Collins, Colorado, June 1999, Vol. 2, pp. 2246–2252.

15　J. Meessen, C. Parisot, C. Lebarzb, D. Nicholsonb, and J.F. Delaigle. WCAM: smart encoding for wireless video surveillance. *Image and Video Communications and Processing 2005*, Proceedings of SPIE Vol. #5685, *IS&T/SPIE 17th Annual Symposium Electronic Imaging*, 16–20 January 2005.

16　T.H. Chalidabhongse, K. Kim, D. Harwood, and L. Davis. A perturbation method for evaluating background subtraction algorithms. In *Joint IEEE International Workshop on Visual Surveillance and Performance Evaluation of Tracking and Surveillance (VS-PETS 2003)*, Nice, France, 11–12 October 2003, pp. 10–116.

17　R. Cucchiara, C. Grana, A. Prati, and R. Vezzani. Using computer vision techniques for dangerous situation detection in domotics applications. In *Second Symposium on Intelligent Distributed Surveillance Systems*, IEE, London, February 2004, pp. 1–5.

18　I.J. Cox and S.L. Hingorani. An efficient implementation of Reid's multiple hypothesis tracking algorithm and its evaluation for the purpose of visual tracking. *IEEE Transactions on Pattern Analysis and Machine Intelligence*, 1996;18(2):138–150.

19　J.H. Piater, S. Richetto, and J. L. Crowley. Event-based activity analysis in live video using a generic object tracker. *Proceedings of the 3rd IEEE International Workshop on PETS*, Copenhagen, 1 June 2002, pp. 1–8.

20　A.E.C. Pece. From cluster tracking to people counting. In *Proceedings of the 3rd IEEE International Workshop on PETS*, Institute of Computer Science, University of Copenhagen, Copenhagen, 1 June 2002, pp. 9–17.

21　B.D. Lucas and T. Kanade. An iterative image registration technique with an application to stereo vision. In *International Joint Conference on Artificial Intelligence*, 1981, pp. 674–679.

22　J. Shi and C. Tomasi. Good features to track. In *IEEE Conference on Computer Vision and Pattern Recognition*, 1994, pp. 593–600.

23　D. Comaniciu, V. Ramesh, and P. Meer. Real-time tracking of non-rigid objects using mean shift. In *IEEE Conference on Computer Vision and Pattern Recognition (CVPR'00)*, Hilton Head Island, South Carolina, 2000, Vol. 2, pp. 142–149.

24　A. Doucet, N. De Freitas, and N.J. Gordon (Eds.). *Sequential Monte Carlo Methods in Practice*. Series Statistics for Engineering and Information Science, 2001, 620 pp.

25　M. Isard and A. Blake. Contour tracking by stochastic propagation of conditional density. In *Proceedings of European Conference on Computer Vision*, Cambridge UK, 1996, Vol. 1, pp. 343–356.

26　K. Nummiaro, E. Koller-Meier, and L. Van Gool. An adaptive color-based particle filter. *Image and Vision Computing*, 2003;21(1):99–110.

27　S. Zhou, R. Chellappa, and B. Moghaddam. Visual tracking and recognition using appearance-adaptive models in particle filters. *IEEE Transactions on Image Processing (IP)*, 2004;11:1434–1456.

28　R. Nevatia, J. Hobbs, and B. Bolles. An ontology for video event representation. In *Conference on Computer Vision and Pattern Recognition Workshop (CVPRW'04)*, June 27–July 2, 2004, Vol. 7, p. 119.

29 X. Desurmont, R. Wijnhoven, E. Jaspert *et al.* Performance evaluation of real-time video content analysis systems in the CANDELA project. In *Conference on Real-Time Imaging IX*, part of the *IS&T/SPIE Symposium on Electronic Imaging 2005*, 16–20 January 2005, San Jose, CA, USA.

30 P.L. Correia and F. Pereira. Objective evaluation of video segmentation quality. *IEEE Transactions on Image Processing*, 2003;12(2):186–200.

31 J.R. Renno, J. Orwell, and G.A. Jones. Evaluation of shadow classification techniques for object detection and tracking. In *IEEE International Conference on Image Processing*, Suntec City, Singapore, October 2004.

32 A. Prati, I. Mikic, M.M. Trivedi, and R. Cucchiara. Detecting moving shadows: algorithms and evaluation. *IEEE Transactions on Pattern Analysis and Machine Intelligence*, 2003;27(7):918–923.

33 P.L. Rosin and E. Ioannidis. Evaluation of global image thresholding for change detection. *Pattern Recognition Letters*, 2003;24(14):2345–2356.

34 Y.J. Zhang. A survey on evaluation methods for image segmentation. *Pattern Recognition*, 1996;29(8):1335–1346.

35 F. Oberti, A. Teschioni, and C.S. Regazzoni. ROC curves for performance evaluation of video sequences processing systems for surveillance applications. In *Proceedings of the International Conference on Image Processing, ICIP 99*, Kobe, Japan, October 1999, Vol. 2, pp. 949–953.

36 F. Oberto, F. Granelli, and C.S. Regazzoni. Minimax based regulation of change detection threshold in video surveillance systems. In G.L. Foresti, P. Mähönen, and C.S. Regazzoni (Eds.), Multimedia Video-Based Surveillance Systems. Kluwer Academic Publishers, 2000, pp. 210–233.

37 T.H. Chalidabhongse, K. Kim, D. Harwood, and L. Davis. A perturbation method for evaluating background subtraction algorithms. In *Proceedings of the Joint IEEE International Workshop on Visual Surveillance and Performance Evaluation of Tracking and Surveillance (VS-PETS 2003)*, Nice, France, October 2003.

38 X. Gao, T.E. Boult, F. Coetzee, and V. Ramesh. Error analysis of background adaption. In *Proceedings of the IEEE Conference on Computer Vision and Pattern Recognition*, Hilton Head Island, SC, USA, June 2000, Vol. 1, pp. 503–510.

39 F. Oberti, E. Stringa, and G. Vernazza. Performance evaluation criterion for characterizing video surveillance systems. *Real-Time Imaging*, 2001;7(5):457–471.

40 V.Y. Mariano *et al.* Performance evaluation of object detection algorithms. *Proceedings of the 16th International Conference on Pattern Recognition*, August 2002, Vol. 3, pp. 965–969.

41 J. Nascimento and J.S. Marques. New performance evaluation metrics for object detection algorithms. In *6th International Workshop on Performance Evaluation for Tracking and Surveillance (PETS 2004)*, ECCV, Prague, Czech Republic, May 2004.

42 C.J. Needham and D. Boyle. Performance evaluation metrics and statistics for positional tracker evaluation. In *Proceedings of the Computer Vision Systems: Third International Conference, ICVS 2003*, Graz, Austria, April 2003, Vol. 2626, pp. 278–289.

43 S. Pingali and J. Segen. Performance evaluation of people tracking systems. In *Proceedings of the 3rd IEEE Workshop on Applications of Computer Vision, 1996, WACV '96*, Sarasota, FL, USA, December 1996, pp. 33–38.

44 M. Rossi and A. Bozzoli. Tracking and counting moving people. In *Proceedings of IEEE International Conference on Image Processing, 1994, ICIP-94*, Austin, TX, USA, November 13–16, 1994, Vol. 3, pp. 212–216.

45 C.E. Erdem, B. Sankur, and A.M. Tekalp. Performance measures for video object segmentation and tracking. *IEEE Transactions on Image Processing*, 2004;13(7):937–951.

46 M. Xu and T. Ellis. Partial observation vs. blind tracking through occlusion. In *British Machine Vision Conference 2002 (BMVC2002)*, University of Cardiff, UK, 2–5 September 2002.

Chapter 6

Tracking objects across uncalibrated, arbitrary topology camera networks

R. Bowden, A. Gilbert and P. KaewTraKulPong

6.1   Introduction

Intelligent visual surveillance is an important application area for computer vision. In situations where networks of hundreds of cameras are used to cover a wide area, the obvious limitation becomes the users' ability to manage such vast amounts of information. For this reason, automated tools that can generalise about activities or track objects are important to the operator. Key to the users' requirements is the ability to track objects across (spatially separated) camera scenes. However, extensive geometric knowledge about the site and camera position is typically required. Such an explicit mapping from camera to world is infeasible for large installations as it requires that the operator know which camera to switch to when an object disappears. To further compound the problem the installation costs of CCTV systems outweigh those of the hardware. This means that geometric constraints or any form of calibration (such as that which might be used with epipolar constraints) is simply not realistic for a real world installation. The algorithms cannot afford to dictate to the installer. This work attempts to address this problem and outlines a method to allow objects to be related and tracked across cameras without any explicit calibration, be it geometric or colour.

Algorithms for tracking multiple objects through occlusion normally perform well for relatively simple scenes where occlusions by static objects and perspective effects are not severe. For example, Figure 6.1 shows a typical surveillance camera view with two distinct regions A and B formed by the presence of a static foreground object (tree) that obscures the ground from view.

(a)　　　　　　　　　　　(b)　　　　　　　　　　　(c)

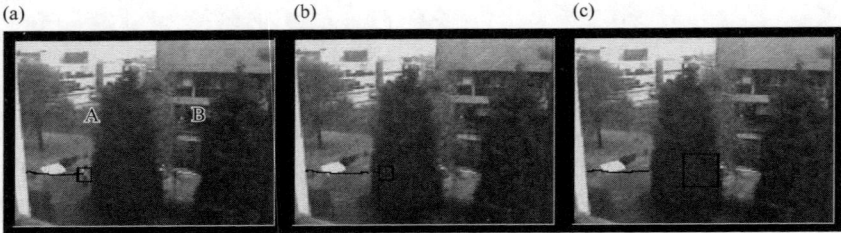

Figure 6.1　Tracking reappearance targets: (a) a target is being tracked; (b) the target is occluded but the search continues; (c) uncertainty propagation increases over time

In this work, regions or sub-regions are defined as separated portions grouped spatially within an image. A region can contain one or more paths which may cover an arbitrary number of areas. Tracking an object across regions in the scene, where the geometric relationship of the regions cannot be assumed, possesses similar challenges to tracking an object across spatially-separated camera scenes (again where the geometric relationships among the cameras are unknown). In a single view, the simplest solution to this problem is to increase the allowable number of consecutive frames that a target persists with no observation before tracking is terminated. This process is shown in Figure 6.1 using a Kalman filter as a linear estimator.

By delaying the tracking termination, both the deterministic and the random components within the dynamics of the Kalman filter propagate over time, increasing the uncertainty of the predicted area in which the target may reappear (as shown in Figure 6.1(c)). This increases the chance of matching targets undergoing long occlusions but also increases the chance of false matches. In situations where linear prediction cannot be assumed (e.g. the pathway changes direction behind large static objects), this will result in a model mismatch, and the kinematic model assumed in most trackers will provide incorrect predictions. Furthermore this type of approach cannot be extended to multiple cameras without an explicit calibration of those cameras.

6.2　Previous work

An approach often used to tackle the tracking of an object across multiple cameras is to ensure some overlap within the field of view of cameras is available. An example is the pilot military system reported in References 1 and 2, where a birds-eye camera view is used to provide a global map. The registration of a number of ground-based cameras to the global map allows tracking to be performed across spatially separated camera scenes. The self-calibrated multi-camera system demonstrated in Reference 3 assumes partially overlapping cameras. Epipolar geometry, landmarks and a target's visual appearance are used to facilitate the tracking of multiple targets across cameras and to resolve occlusions.

Tracking across non-overlapping views has been of recent interest to many researchers. Huang and Russell [5] developed a system to track vehicles on a highway across spatially separated cameras. In their system, knowledge about the entrance/exit of the vehicles into the camera views must be provided along with transition probabilities. Kettnaker and Zabih [6] presented a Bayesian framework to track objects across camera views. The user supplies topological knowledge of usual paths and transition probabilities. Javed *et al.* [7] present a more general solution to the problem by providing an update of inter-camera parameters and appearance probabilities. However, their method assumes initial correspondences of those parameters. Our method makes use of data obtained automatically to discover such relationships between camera views without user intervention.

In accumulating evidence of patterns over time we expect to discover common activities. These patterns can be modelled in a number of ways. They can be used to classify sequences as well as individual instances of a sequence or to discover common activities [8]. Howarth and Buxton [9] introduce a spatial model in the form of a hierarchical structure of small areas for event detection in traffic surveillance. The model is constructed manually from tracking data. Fernyhough *et al.* [10] use tracked data to build a spatial model to represent areas, paths and regions in space. The model is constructed using a frequency distribution collected from a convex-hull binary image of the objects. Thresholding of the frequency distribution filters out low distribution areas, that is, noise. Johnson and Hogg [11] use flow vectors, that is, the 2D position and velocity, collected from an image sequence over an extended period to build a probability distribution of the targets moving in the image. A neural network with competitive layers is used to quantise the data and represent the distribution. Its use is to detect atypical events which occurred in the scene. Makris and Ellis [12] use spline representations to model common routes and the activity of these routes. Entry–exit points and junctions are identified as well as their frequencies. Nair and Clark [13] use extended data to train two HMMs to recognise people entering and exiting a room in a corridor. Uncommon activity is identified as break-in by calculating the likelihood of the trajectories of the HMMs and comparing with a pre-defined threshold. Stauffer [14] uses on-line vector quantisation as described in Reference 11 to quantise all target information including position, velocity, size and binary object silhouette into 400 prototypes. Then they perform inference on their data to obtain a probability distribution of the prototypes encoded in a co-occurrence matrix. A normalised cut [15] is then performed on the matrix which results in grouping similar targets in terms of the above features in a hierarchical form.

6.3 Overview

Figure 6.2 gives a general overview of the system with the per camera elements separated from those that correlate objects between cameras and regions. Each camera is first fed to an object detection module, described in Section 6.4. Here a background scene model is maintained and used to segment foreground objects on the fly. This basic segmentation is then further refined by identifying areas of misclassification

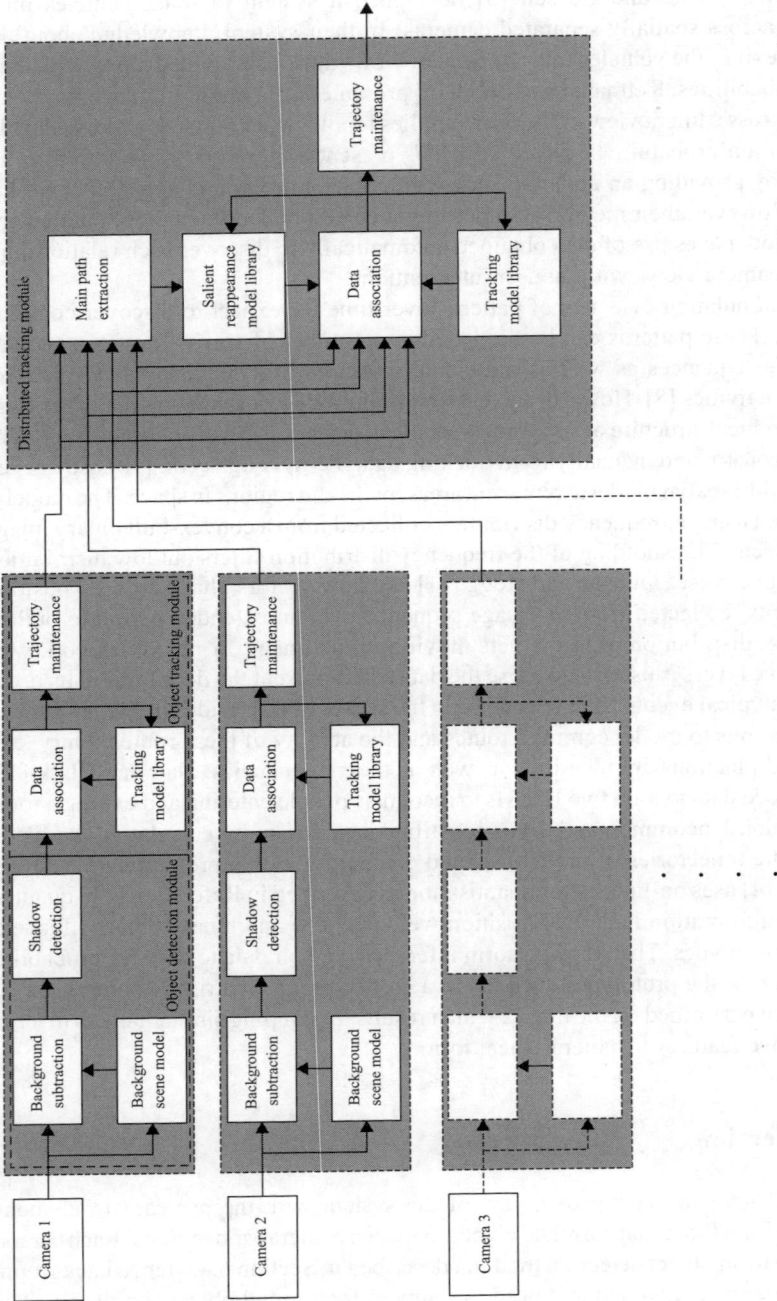

Figure 6.2   System overview

due to shadows using the chromaticity of individual pixels. Following this, objects are passed to the object tracking module (Section 6.5), where data association attempts to provide consistent temporal labelling of objects as they move. This is done by maintaining a library of currently tracked objects which summarises all measurements into the relevant motion and appearance models. The outputs of this module are trajectories that exhibit temporal and spatial consistency.

Following this, the resulting tracked objects in each camera are passed to the distributed tracking module (Section 6.7), which attempts to connect seemingly unrelated trajectories (both spatially and temporally) into consistent object labels across regions and cameras, the first step of which is to extract main paths. This involves an unsupervised clustering of trajectories: grouping like trajectories and discarding spurious detections and any trajectories which have insufficient supporting evidence to provide reliable results. Following this a model library of salient reappearance periods is constructed between all main paths and used in the data association stage to link trajectories that are spatially and temporally distinct from each other. The output of this module is the consistent labelling of objects despite occlusions or disappearances between cameras.

6.4   Object detection module

The module for object detection consists of three parts. First, each pixel in the input image is segmented into moving regions by a background subtraction method using a per pixel mixture of Gaussians as the reference image. This base segmentation is then fed into a shadow detection module to eliminate shadows from moving objects. The resulting binary image is then grouped into different objects by the foreground region detection module.

6.4.1   Background modelling

The model operates within a framework similar to that introduced by Stauffer and Grimson [8]. The difference lies in the update equations of the model parameters and the initial weight of a new Gaussian component (explained shortly). In previous work we have demonstrated the superior performance of update equations derived from sufficient statistics and the L-recent window formula over other approaches [20,21]. The derivation of the update equations is given in Reference 16. This provides a system which learns a stable background scene faster and more accurately than do other approaches.

Each pixel in the scene is modelled by a mixture of K Gaussian distributions (K is a small number from 3 to 5). Different Gaussians are assumed to represent different colours. The probability that a certain pixel has a value of $\mathbf{x}_N$ at frame N can be written as

$$p(\mathbf{x}_N) = \sum_{j=1}^{K} \omega_j \eta(\mathbf{x}_N; \mu_j, \Sigma_j), \tag{6.1}$$

where ω_k is the weight parameter of the kth Gaussian component which represents the time proportions that the colour stays in the scene,

$$\sum_{j=1}^{K} \omega_j = 1 \quad \text{and} \quad \forall_j, \quad \omega_j \geq 0. \tag{6.2}$$

$\eta(\mathbf{x}; \mu_k, \Sigma_k)$ is the Gaussian distribution of the kth component, where μ_k is the mean and $\Sigma_k = \sigma_k^2 \mathbf{I}_d$ is the covariance of the kth component. d is the dimensionality of the vector $\mathbf{x} \in \Re^d$. This simplification reduces model accuracy but provides a significant increase in efficiency as it removes the need for matrix inversions on a per pixel level.

The background components are determined by assuming that the background contains the B most probable colours. These probable background colours are the ones that remain static for a large portion of time. To identify background components, the K distributions are ordered based upon their fitness ω_k/σ_k, and the first B distributions are determined by

$$B = \arg\min \left(\sum_{j=1}^{b} \omega_j > \text{th} \right). \tag{6.3}$$

$\{\omega_1, \omega_2, \ldots, \omega_k\}$ are now the weight parameters of the mixture components in descending orders of fitness. The threshold th is the minimum fraction of the model that is background. In other words, it is the minimum prior probability that the background is in the scene, that is, th $= 0.8$ is the prior probability that 80 per cent of the scene variation is due to static background processes. Background subtraction is performed by marking any pixel that is more than 2.5 standard deviations away from all B distributions as a foreground pixel; otherwise a background pixel.

If the above process identifies any match to the existing model, the first Gaussian component that matches the test value will be updated with the new observation by the update equations,

$$\hat{\omega}_k^{(N+1)} = \hat{\omega}_k^{(N)} + \alpha^{(N+1)} \left(M_k^{(N+1)} - \hat{\omega}_k^{(N)} \right),$$

$$\hat{\mu}_k^{(N+1)} = \hat{\mu}_k^{(N)} + \rho^{(N+1)} \left(\mathbf{x}_{N+1} - \hat{\mu}_k^{(N)} \right), \tag{6.4}$$

$$\hat{\Sigma}_k^{(N+1)} = \hat{\Sigma}_k^{(N)} + \rho^{(N+1)} \left(\left(\mathbf{x}_{N+1} - \hat{\mu}_k^{(N)} \right) \left(\mathbf{x}_{N+1} - \hat{\mu}_k^{(N)} \right)^T - \hat{\Sigma}_k^{(N)} \right),$$

where

$$\alpha^{(N+1)} = \max \left(\frac{1}{N+1}, L \right) \tag{6.5}$$

and

$$\rho^{(N+1)} = \max \left(\frac{1}{\sum_{i=1}^{n+1} M_i^{(N+1)}}, \frac{1}{L} \right). \tag{6.6}$$

The membership function which attributes new observations to a model component $M_k^{(t+1)}$ is set to 1 if ω_k is the first matched Gaussian component; 0 otherwise.

Here ω_k is the weight of the kth Gaussian component, $\alpha^{(N+1)}$ is the learning rate and $1/\alpha$ defines the time constant which determines change. N is the number of updates since system initialisation. Equations (6.4)–(6.6) are an approximation of those derived from sufficient statistics and the L-recent window to reduce computational complexity [16].

If no match is found to the existing components, a new component is added. If the maximum number of components has been exceeded, the component with lowest fitness value is replaced (and therefore, the number of updates to this component is removed from N). The initial weight of this new component is set to $\alpha^{(N+1)}$ and the initial standard deviation is assigned to that of the camera noise. This update scheme allows the model to adapt to changes in illumination and runs in real-time.

6.4.2   Shadow elimination

In order to identify and remove moving shadows, we need to consider a colour model that can separate chromatic and brightness components. It should also be compatible and make use of our mixture model. This can be done by comparing non-background pixels against the current background components. If the differences in both chromatic and brightness components are within some threshold, the pixel is considered as a moving shadow. We use an effective computational colour model similar to the one proposed by Horprasert *et al.* [22] to meet these needs. It consists of a position vector at the RGB mean of the pixel background, $\mathbf{E}$, an expected chromaticity line, $\|\mathbf{E}\|$, a chromatic distortion, d, and a brightness threshold, τ. For a given observed pixel value, $\mathbf{I}$, a brightness distortion a and a colour distortion c, the background model can be calculated by

$$a = \arg\min_z (\mathbf{I} - z\mathbf{E})^2 \tag{6.7}$$

and

$$c = \|\mathbf{I} - a\mathbf{E}\|. \tag{6.8}$$

With the assumption of a spherical Gaussian distribution in each mixture component, the standard deviation of the kth component σ_k can be set equal to d. The calculation of a and c is trivial using a vector dot product. A non-background observed sample is considered a moving shadow if a is within, in our case, 2.5 standard deviations and $\tau < c < 1$.

6.5   Object tracking module

The object tracking module deals with assigning foreground objects detected from the object detection module to models maintained in the target model library within a single camera. It also handles situations such as new targets, targets that are temporarily lost, occluded or camouflaged and targets whose appearance merges with others. This task incorporates all available information to choose the best hypothesis to match. The process consists of data association, stochastic sampling search and the trajectory maintenance modules.

6.5.1   Target model

In the system, multiple objects are tracked based on information about their position, motion, simple shape features and colour. The characteristics of an object are assumed to be independent, as no constraints are placed upon the types of objects that can be tracked. Therefore, separate models are employed for each attribute.

A discrete time kinematic model models the co-ordinates of the object's centroid. Kalman filters are employed to maintain the state of the object using a white noise acceleration term. This follows the assumption that objects move with a near constant velocity.

Shape is represented by the height and width of the minimum bounding box of the object. An excessive change of the shape/size from an average size indicates an object may be under camouflage or partial occlusion. The average size estimate, $\hat{h}$, is maintained using an update equation similar to that of $\mu_k^{(N+1)}$ in Equation (6.4).

Colour is represented as consensus colours in Munsell colour space by converting observed colours into 11 basic colours [23]. This consensus colour was experimentally developed by Sturges and Whitfield [17]. The colour conversion is obtained by assigning colours to consensus regions in a nearest neighbour sense. A colour histogram containing 11 normalised bins is then built and updated in a manner similar to that of $\hat{h}$, using the previously described update equations.

An example of the result from the conversion is shown in Figure 6.3. In this figure, only foreground pixels are converted into consensus colours via a lookup table. A more detailed discussion is presented in Section 6.6.

6.5.2   Data association module

This module makes use of both the motion and appearance models of the targets. Each tracking model includes motion and appearance. The motion model gives an ellipsoidal prediction area called the validation gate [4]. This area is represented by a squared Mahalanobis distance less than or equal to a gate threshold from the predicted measurement with a covariance matrix being the Kalman innovation covariance matrix **S**. The squared Mahalanobis distance is a chi-squared distributed with the number of degrees of freedom equal to the dimension of the measurement vector. Hence, the probability of finding the measurement in the gate, that is, having the Mahalanobis distance less than the gate threshold, can be obtained from the chi-squared distribution.

As targets come close to each other, a measurement from one target may fall within more than one validation gate and an optimal assignment is then sought. The purpose of data association is to assign the measurements detected in the current frame to the correct target models. Targets whose parts cannot be detected due to failure of the background segmentation are deemed to be camouflaged and therefore

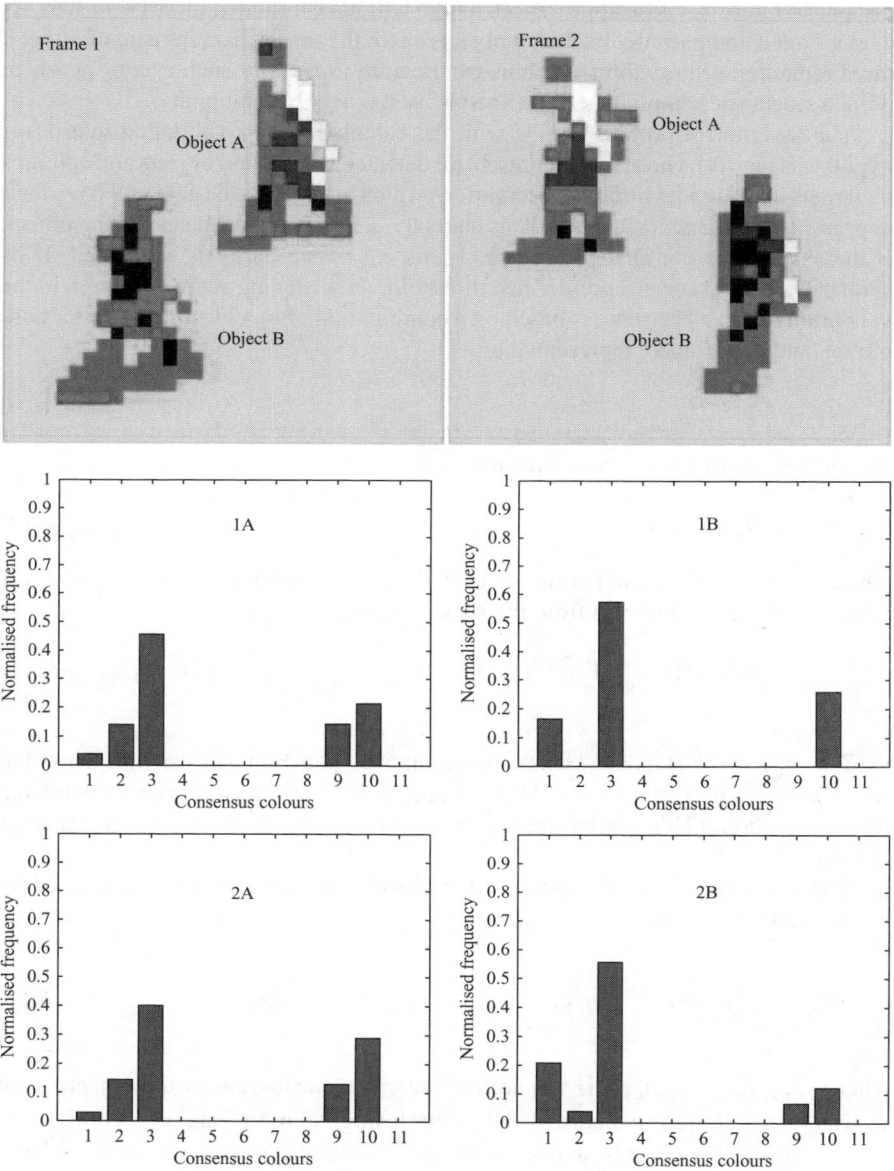

Figure 6.3   Example colour descriptors for two objects over time

disappear. Objects whose appearances merge will have a sudden increase in size. As camouflaged and partially occluded objects share the same characteristic of either a rapid reduction or growth in size, this can be used to identify such events, at which point a stochastic sampling search (SSS) is used to resolve ambiguity.

The assignment process begins with the calculation of a validation matrix (or hypothesis matrix) whose rows represent all detected foreground objects and columns all targets in the model library. Observations which are within this gate will have their appearance similarity calculated. This starts by determining the shape of the object. If the shape does not change extensively, its colour similarity is calculated. If its similarity score exceeds a pre-defined threshold, the matching score is entered in the validation matrix. The score is based on a combination of model similarities for both motion and appearance, represented by

$$T_{ij} = M_{ij} + H_{ij}. \tag{6.9}$$

The motion score M_{ij} is represented by

$$M_{ij} = \Pr(Z > z_{ij}), \tag{6.10}$$

where z_{ij} is the Mahalanobis distance of the ith measurement ($z_i(k + 1)$) to the estimated position predicted from the jth target ($\hat{z}_j(k + 1|k)$).

$$z_{ij} = ((z_i(k + 1) - \hat{z}_j(k + 1|k))^T S_j(k + 1)^{-1} (z_i(k + 1) - \hat{z}_j(k + 1|k)))^{1/2}. \tag{6.11}$$

$\Pr(Z > z_{ij})$ is the standard Gaussian cumulative probability in the right-hand tail which gives the maximum value of 0.5 if the measurement coincides with the predicted (mean) location. (This can be implemented in a look-up table to increase speed of operation.)

The colour similarity H_{ij} between the object i and the target j is calculated by histogram intersection,

$$H_{ij} = \frac{1}{2} \sum_{k=1}^{11} \min(B_{i,k}, \hat{B}_{j,k}), \tag{6.12}$$

where $\{B_{i,1}, B_{i,2}, \ldots, B_{i,11}\}$ is the normalised colour histogram of the object i, and $\{\hat{B}_{j,1}, \hat{B}_{j,2}, \ldots, \hat{B}_{j,11}\}$ is the normalised colour histogram of the target j.

The best solution can be defined as the assignment that maximises the hypothesis score T_{ij} over all possible matches. As this assignment is crisp (not fuzzy), one solution is to modify the existing validation matrix by adding new hypothesised targets or undetected measurements to form a square matrix and run an assignment algorithm such as the Hungarian Algorithm. After the data association process, all assigned measurements are removed from the binary map. The binary map is then passed to the stochastic sampling search (SSS) to extract measurement residuals available for the unassigned targets.

6.5.3 Stochastic sampling search

If camouflage and occlusions occur, measurements of some observations may not be assigned a target. All unassigned targets are then passed to the SSS along with the binary map obtained from the object detection module with all assigned measurements removed. The SSS is a method that incorporates measurement extraction, motion tracking and data association in the same process.

It begins by sorting the unassigned target models according to some depth estimate. This can be obtained through modelling of the ground plane or in the simple case using the y co-ordinate of the object (assuming that targets move on a ground plane in perspective view). A number of patches with approximately the same size as the target are generated. The locations of these patches are randomly sampled from the probability density function (pdf) of the motion model S. In each patch, only the pixels marked by the binary image are considered and converted to consensus colours and a colour histogram constructed.

As this patch may include pixels generated from other targets, normal histogram intersection would not give optimal results. Instead of normalising the histogram, each bin in the histogram is divided by the estimated number of pixels in the target before the colour similarity is calculated. The colour similarity and motion score of each patch are then calculated as done previously and the optimum patch chosen as that which maximises the model similarity score. This estimate can be used to update the motion model (however, not the appearance model), provided that the similarity of the patch exceeds some pre-defined threshold. This threshold is the approximate percentage of the visible area of the target.

The matched patch is removed from the binary map. The binary map is then passed to the track maintenance module to identify new targets.

6.5.4 Trajectory maintenance module

The track maintenance module is designed to deal with trajectory formation and trajectory deletion as well as to eliminate spurious trajectories that occur from unpredicted situations in outdoor scenes, such as trajectories resulting from noise and small repetitive motions. By performing connected component analysis on the residual binary image, a list of new objects which have a suitable number of pixels is extracted. This provides evidence of all objects not already accounted for by the tracking system. Track formation is described as follows. First, every unassigned measurement is used to form a track, called a 'tentative' object. At the next frame, a gate estimate is formed by propagating the process and measurement uncertainties from the last position of the target. If a measurement is detected in the gate, this tentative track becomes a 'normal' track; otherwise it is discarded. The construction of an appearance model for the new target is deferred (as the object's initial appearance is normally unstable). The tracking process during this period relies solely on the motion model. If no measurement has been assigned to a normal track it is then changed to a 'lost' object. If a normal track is not assigned a measurement during the data association process, it is changed to 'occluded'. Any type of track can be changed back to normal if it is assigned a measurement during the data association process. Tracks that have been lost for a certain number of frames are deleted; this also applies to occluded tracks.

6.6   Colour similarity

A similarity metric based upon histogram intersection provides a crude estimate of object similarity. Through quantisation, it also allows some invariance to changes in colour appearance. Three colour spaces were investigated, RGB, HSL and consensus-colour conversion of Munsell colour space (termed CLUT below for colour look-up table) as proposed by Sturges and Whitfield [17]. A series of tests were performed to investigate the colour consistency of these methods for objects moving within a single image and also across images. CLUT breaks all colours down into 11 basic colours; the exact membership was determined experimentally in a physiological study where human categorisation was used to learn the membership. This coarse quantisation should provide consistent intra-/inter-camera labelling without colour calibration relying on the perceptual consistency of colour; that is, if a red object is perceived as red in both images CLUT will provide a consistent label. For CLUT, the quantisation into 11 pre-defined bins is fixed; however, for both RGB and HSL the level of quantisation must be selected. Three different quantisation levels were investigated for RGB: three bins per colour channel, resulting in 27 bins; four bins per colour channel resulting in 64 bins; and five bins per channel resulting in a 125 bin histogram. HSL was quantised using 8–8–4 bins across the respective channels H–S–L (as suggested in Reference 18).

The initial experiment looked at the performance of the various colour descriptors for intra-camera tracking, that is, given the repeated occurrence of an object, which colour descriptor provides the most reliable match within a single camera. The results of this test are shown in Table 6.1. Ten people were tracked and manual ground truth obtained. The mean histogram intersection and variance were calculated across all occurrences of a person. The results show that there is little difference between the various colour spaces and levels of quantisation. HSL colour space provides marginally better results over other spaces. This is not unsurprising as its perceptual separation of chromatic and luminosity should make it less susceptible to variations in ambient lighting.

However, as the colour descriptor must not only perform intra-camera object tracking but allow correlation of objects across cameras (inter-camera), the consistency of the descriptors at matching the same objects between cameras must be established. It is also important to consider how well the colour descriptors can discriminate between objects which do not match. To this end a second set of tests were performed where ten objects were tracked across four non-overlapping cameras with no colour calibration. Manual ground truth was established and the results are shown in Table 6.2. The 'Matched' results show the mean (and standard deviation in brackets) of objects correctly matched to themselves. Here a low score with correspondingly low variance (to show consistency) is desirable. The scores are an order of magnitude higher than those of Table 6.1; this is due to the lack of colour

Table 6.1   *Average histogram intersection for various colour spaces and quantisation levels within a single camera image*

	CLUT	RGB			HSL
	11	$3 \times 3 \times 3$	$4 \times 4 \times 4$	$5 \times 5 \times 5$	$8 \times 8 \times 4$
Mean histogram intersection	0.0598	0.0544	0.0621	0.0549	0.0534
Variance	0.0211	0.0219	0.0167	0.0127	0.0176

Table 6.2   *Average histogram intersection for both correct and falsely matched objects for varying colour spaces and quantisation levels across cameras*

	CLUT	RGB			HSL
	11	$3 \times 3 \times 3$	$4 \times 4 \times 4$	$5 \times 5 \times 5$	$8 \times 8 \times 4$
Matched (same person)	0.24(0.10)	0.19(0.13)	0.24(0.11)	0.28(0.14)	0.37(0.16)
Non-matched (different person)	0.58(0.24)	0.37(0.20)	0.41(0.17)	0.42(0.16)	0.64(0.14)
T test	2.19	1.41	1.45	1.11	1.92

calibration between cameras. In Table 6.1 the single camera means that it is effectively perfectly colour calibrated with itself and hence the much lower scores. For inter-camera tracking without colour consistency, HSL now performs the worst, giving the highest match. However, it is the colour space's ability to distinguish between objects that do not match which should also be considered. The 'Non-matched' results show the mean (and standard deviation) of scores between unrelated objects across cameras. Here a high mean with low variance is desirable. To assess the suitability of each of the colour spaces to inter-camera correspondence a statistical T test was performed to highlight the colour space which produced the most distinct difference between matched and non-matched object correspondence. Here it can clearly be seen that the CLUT approach gives superior performance with HSL a close second. Not surprisingly, RGB performs poorly regardless of the level of quantisation used.

CLUT provides superior results inter-camera while retaining results comparable to HSL intra-camera. Figure 6.3 shows two images from a sequence labelled frame 1 and frame 2. These frames contain the same two objects (A and B) at different times. Although spatially the appearances of the objects differ, the CLUT colour histograms are relatively consistent for each.

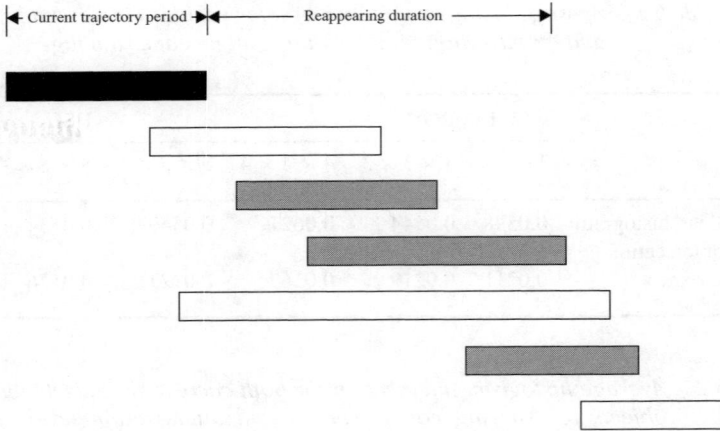

Figure 6.4   An example timeline of plausible matching

6.7   Relating possible reappearing targets

In order to identify the same target reappearing after a long occlusion, features/
characteristics of object similarity must be identified. The properties of reappearing
targets are assumed as follows:

- A target should disappear from one area and reappear in another within a certain
  length of time.
- The reappearing target must occur only after that target has disappeared. (There
  is no co-existence of the same target at any instance. This is according to the
  spatially separated assumption. For overlapped camera views, this assumption
  may not be applied.)
- Frequently occurring disappearances or reappearances will form consistent trends
  within the data.

An example of plausible matching can be seen in Figure 6.4. In this figure the black
bar indicates the time line of the target of interest. Grey bars are the targets that can
make possible matches to the target of interest, whereas white bars represent targets
whose matching with the target of interest are illegal due to breaches of the above
assumptions.

If targets are moving at similar speeds along a path occluded by large static objects
(such as the tree in Figure 6.1), over a period of time, there should be a number of
targets that disappear from a specific area of region A and reappear within another area
of region B. This can be called the salient reappearance period between the two areas.
Both areas can be considered to be elements of the same path even though they
are in different regions. The reappearance periods of these targets should be similar
compared with random appearance of targets between other regions.

Figure 6.5   Trajectory data of the training set

Data are automatically collected from our single camera-tracking algorithm as previously described. The data consist of the tracking details of targets passing into the field of view of the camera. Figure 6.5 shows trajectories of all the objects collated. Note that due to the occlusion of the tree, two separate regions are formed where no correspondence is known about the relationship between them. The linear dynamics of the tracker are insufficient to cope with the occlusion.

The goal is to learn some reappearance relationship in an unsupervised fashion to allow objects which disappear to be successfully located and tracked when they reappear. This is termed the salient reappearance period, and its construction is divided into two steps:

1. Extracting dominant paths: trajectories in each subregion are classified into a number of paths. Only 'main' paths that consist of a large number of supporting trajectories are extracted.
2. Extracting salient reappearance periods among dominant paths: in this stage, a set of features common to reappearing targets are introduced. The features allow possible reappearance among pairs of paths to be calculated. A set of matches that show outstanding reappearance periods are chosen to train the matching models in the training phase.

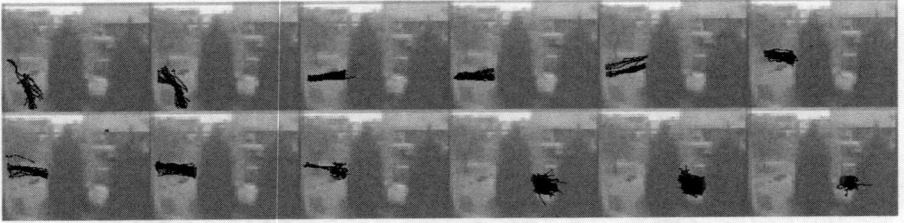

Figure 6.6   Main paths in the first region

6.7.1   Path extraction

Paths can be effectively represented as a group of trajectories between two areas. However, some trajectories may begin and end in the same area. Paths of this type must be divided into different groups. This is done using a normalised cut [15] of the 4D motion vectors formed through the concatenation of position and velocity (x, y, dx, dy). All paths which contain more than 2 per cent of the total trajectories are kept for further processing and are shown in Figure 6.6. Paths with a small number of supporting trajectories are sensitive to noise and can produce erroneous trends. They are therefore discarded. The figure shows how the normalised cut has broken trajectories down into groups of like trajectories.

6.7.2   Linking paths

Paths are linked by building a fuzzy histogram of all possible links among trajectories of a pair of main paths from different regions within a period named the 'allowable reappearance period'. The bin τ_t of the histogram is calculated from

$$\tau_t = \sum_{\forall i,j} H_{ij}; (t_i^{\text{end}} - t_j^{\text{start}}) < t, \tag{6.13}$$

where t_i^{start} and t_i^{end} are the time instances that target i starts and ends, respectively. H_{ij} is the result from histogram intersection between the colour histogram of target i and that of target j.

An example of the fuzzy histogram of reappearance periods within 60 s from sample path A to B in Figure 6.7 is shown in Figure 6.8. The histogram bin size was set at 1 s, and the frequency of the bin was the summation of colour similarity scores using CLUT. For a full discussion of the allowable reappearance period, the fuzzy frequency and the choice of parameters used, the interested reader is pointed to Reference 19.

Using the linking scheme described previously on every pair of main paths between two regions (with only main paths from different regions permitted for the matches), a cumulative number of possible matches produces salient reappearance periods. Figure 6.9 shows the results. Two histograms for every pair of paths are produced. The one that has the maximum peak is selected. To check the validity of the

Figure 6.7   An example of a pair of possible paths to be matched

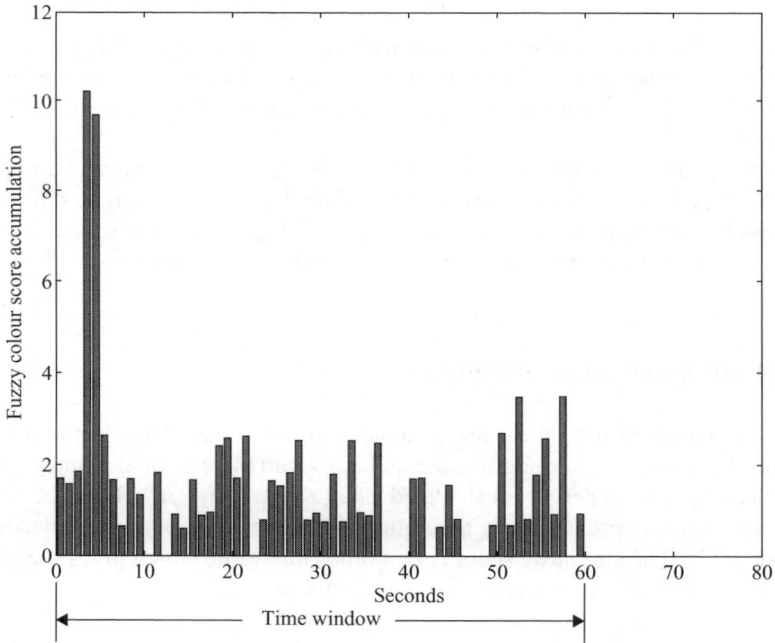

Figure 6.8   Fuzzy histogram of reappearance periods from sample path A to B with an allowable reappearance period of 60 s

outstanding peak it must exceed four times the noise floor level. The noise floor level was found by taking the median from non-empty bins of the histogram. Unimodality is assumed, and a single peak is detected based on the maximum area under the bins that pass the noise level. This could of course be used in a particle filter framework, should unimodality not be sufficient. However, our work thus far has not found it

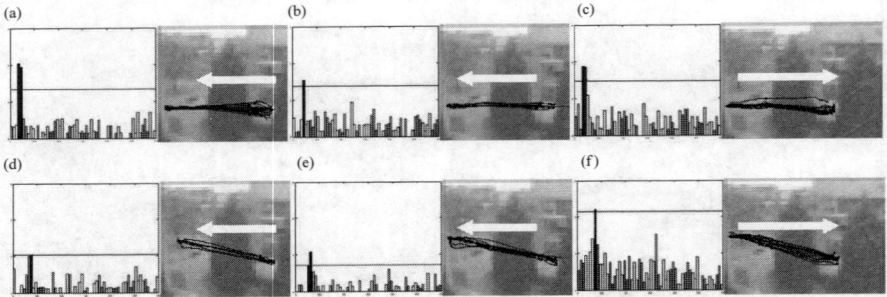

Figure 6.9   Extracted salient reappearance periods among main paths between different regions

necessary. The histograms are presented with the corresponding linked trajectories in Figure 6.9, with white arrows showing the direction of motion. The detected bins are shown with solid bars on each histogram, and the noise level is also plotted in each histogram.

With the data automatically collected from the last process, a reappearing-target model can be formed for each pair of detected main paths. For each pair of detected main paths, the reappearance periods $\{r_1, r_2, \ldots, r_N\}$ between a pair of paths are represented compactly by their mean μ_r and standard deviation σ_r.

6.8   Path-based target recognition

Similarity based on the salient reappearance periods is calculated in the following manner. First, the standard (zero mean and unity variance) normal random variable of each reappearance period is calculated using $z = (r - \mu_r)/\sigma_r$. Then the standard normal cumulative probability in the right-hand tail $\Pr(Z > z)$ is determined using a look-up table. The similarity score is two times this value which gives the score in the range of 0–1. Another way of calculating this value is

$$2\Pr(Z > z) = 1 - \int_0^z f(x|1)\,dx, \qquad (6.14)$$

where $f(x|v)$ is the chi-square distribution with v degree of freedom.

$$f(x|v) = \frac{x^{(v-2)/2}e^{-x/2}}{2^{v/2}\Gamma(v/2)}, \qquad (6.15)$$

where $\Gamma(v)$ is the Gamma function. However, a no match hypothesis or null hypothesis is also introduced, as it is possible to have no link. Hypothesis testing for this null hypothesis is required before any classification is performed. A score of 0.001 was set for the null hypothesis which corresponds to the standard value z of 3.09. Any candidates which are not null hypotheses are selected based on their maximum score. Online recognition selects the best candidate at each time instance within the allowable reappearance period. This allows the tracker to change its link each time a better hypothesis is introduced. Batch recognition on the other hand collects all trajectories within the allowable reappearance period and performs the classification based on the whole set.

In the first experiment, the training data were collected automatically by the target tracking algorithm over an extended period of time constituting 1009 individual trajectories. An unseen test set of 94 trajectories was collected and hand labelled as ground truth. The recognition process is performed off-line by collecting the whole set of admissible targets according to the rules described in Section 6.7. An example of the recognition with similarity score and the corresponding time line chart is shown in Figure 6.10. The motion of the objects is highlighted in the figures, where white arrows denote implausible motions and black arrows non-null (plausible) hypotheses. The red line in the time line is the target of interest, while the green lines are the non-null hypothesis candidates with plausible matches. The arrows depict the ground truth. It can be seen that target (b) was classified correctly to the ground truth as it obtained the best score during the recognition process.

A summary of the recognition on the whole test set is provided in Table 6.3. Since the model is based on the trend in a pair of main paths, if the target uses uncommon paths which have no trend in the dataset, the target cannot be recognised. This accounts for two out of the six misses. One of these misses was caused by a person who changed his mind during his disappearance and walked back to the same path in the same area, while the other was due to camouflage at the beginning of the trajectory.

Table 6.3   Matching results of an unseen dataset in a single camera view

Items	Trajectories	%
Total training set	1009	
Total test set	94	100.00
Correct matches	84	89.36
Total incorrect matches	10	10.64
False detections	4	4.26
Misses	6	6.38
Misses (uncommon paths)	2	2.13

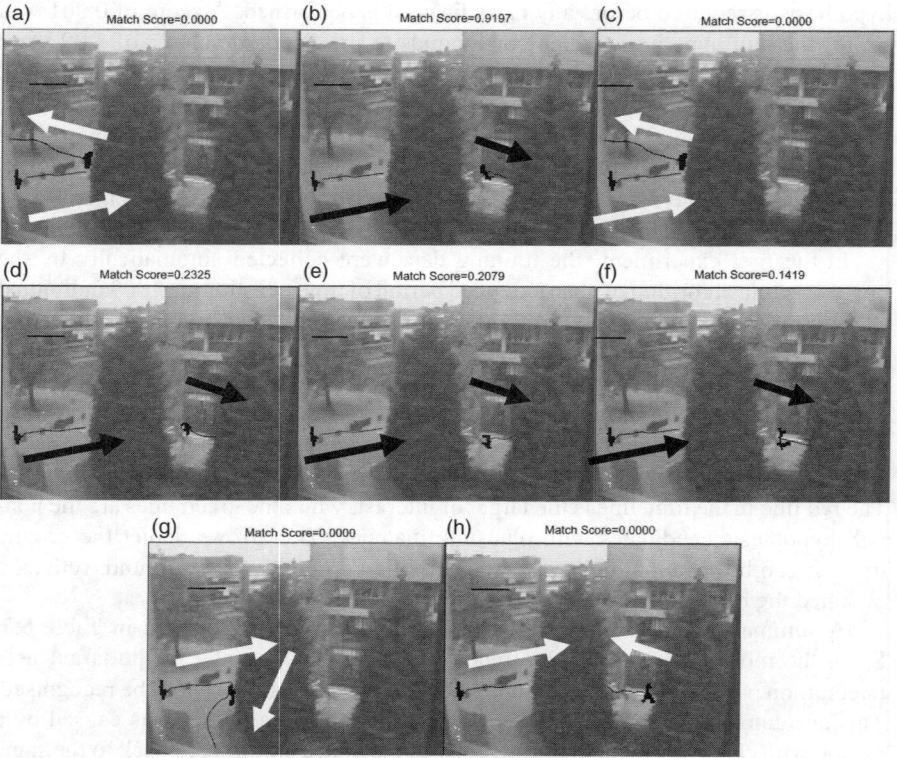

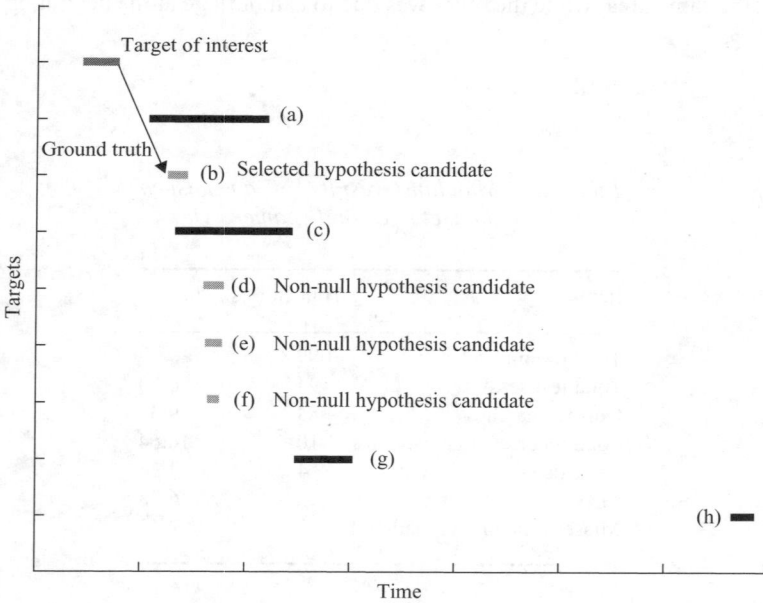

Figure 6.10   An example of recognising a reappearing target between different regions and the associated timeline

6.9   Learning and recognising trajectory patterns across camera views

The target recognition presented thus far can also be extended to multiple cameras. The same process of trajectory data collection from two scenes with time synchronisation was performed. The set-up of cameras is shown in Figure 6.11. A total of 2,020 individual trajectories consisting of 56,391 data points were collected for two cameras. An unseen test set of 133 trajectories was then hand labelled to provide ground truth.

Figure 6.12 shows some examples of the salient appearing periods and trajectories extracted between camera pairs for extracted main paths. Note that both the examples are valid reappearing trends. However, the trend in Figure 6.12(a) is not as high due to a small sample set as well as an increased distance between the two main paths. It should be noted if any two regions are physically too distant, the prominence of the salient reappearance period is reduced.

The results from pre-processing are then subjected to the same process as before and classification performed in the same way as in the single view. The test data were collected and hand labelled. They consist of 133 targets. For an example trajectory, Figure 6.13 shows all possible links to other paths within the time window along with the corresponding time line chart in Figure 6.14. Again arrows assist in visualising the direction of motion, and colour coding is used to depict plausible (black) matches against null hypothesis (white) matches. The two highest candidates are reiterated in Figure 6.15. In this example, candidate (b) is the best match and is selected. The second best, which is candidate (g), is also valid; however, it has a lower score due to the flatter peak in its training set. This higher variance is caused by the greater distance between the two main paths which increases the effect of variance of the

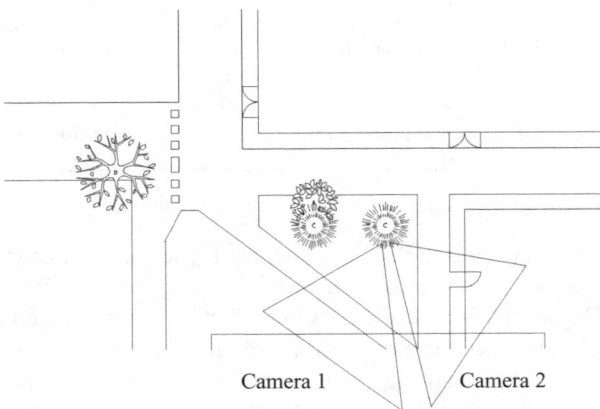

Camera 1   Camera 2

Figure 6.11   Site map and camera layouts for recognising reappearing targets across camera views

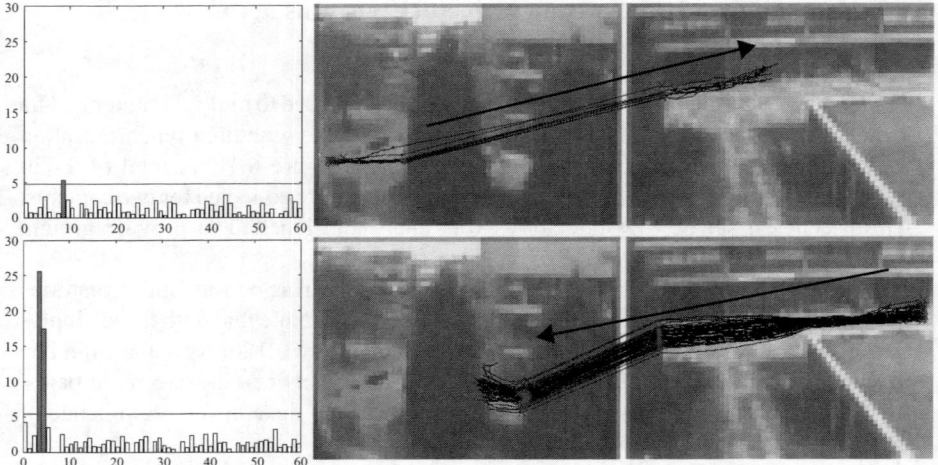

Figure 6.12   Examples of extracting salient reappearance periods between main paths in different regions across camera views

Table 6.4   *Matching results of an unseen set of 10 min across camera views*

Items	Trajectories	%
Total training set	2020	
Total test set	133	100.00
Correct matches	116	87.22
Total incorrect matches	17	12.78
False detections	8	6.02
Misses	9	6.77
Misses (uncommon paths)	2	1.50

target speed. It is interesting to note that both the highest matches are in fact correct as the current object becomes (b) and then after disappearing for a second time becomes (g). The approach naturally tries to predict as far into the future as possible. However, in practice, once the second disappearance has occurred, a much stronger direct match between (b) and (g) would be used to perform the match. Matching results from the two scenes are shown in Table 6.4. Again a high number of correct matches is achieved, only slightly lower than that of the single camera results. Most notable is that using colour alone to match targets results in only around 60 per cent

(a) Match score = 0.0296   (b) Match score = 0.9060

(c) Match score = 0.0000   (d) Match score = 0.1257

(e) Match score = 0.0000   (f) Match score = 0.0000

(g) Match score = 0.7205   (h) Match score = 0.0000

(i) Match score = 0.0000   (j) Match score = 0.0000

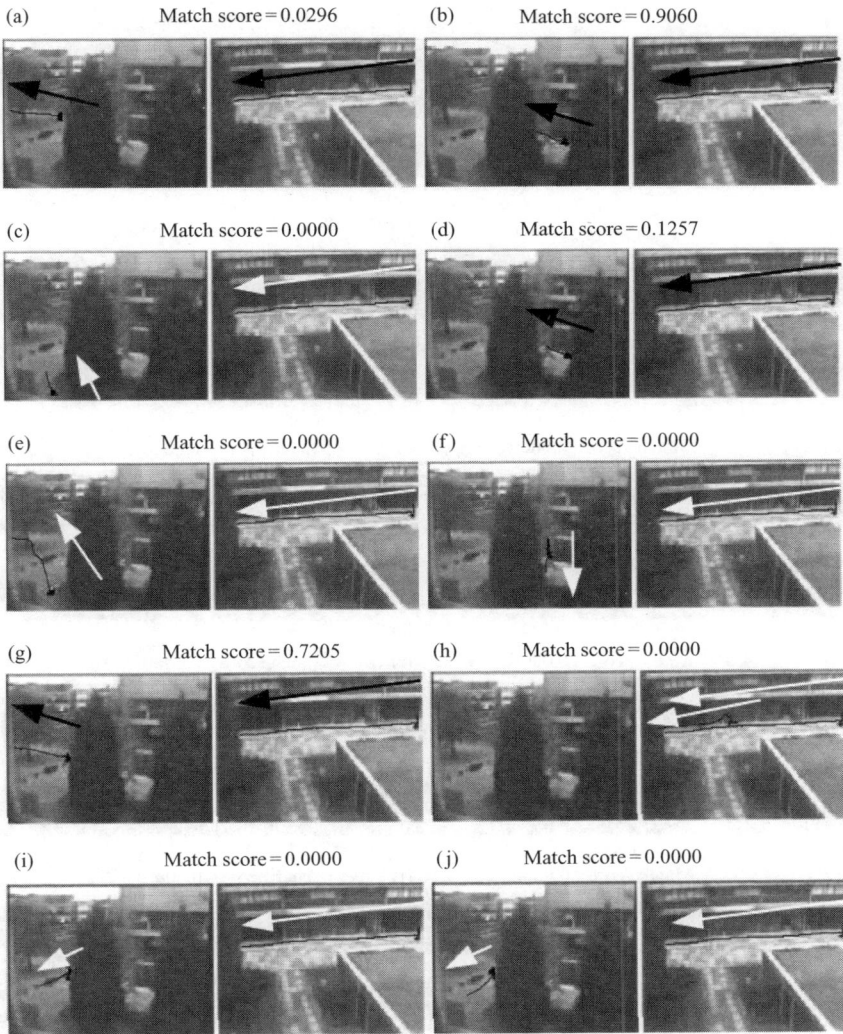

Figure 6.13a   All possible links between paths in separate views

success rate, but using colour to look for statistical trends spatially can achieve over 87 per cent. More important is the number of false detections rather than misses; in terms of system performance and human evaluation it is this that is used to assess performance.

It can be seen from Tables 6.1 and 6.2 that the recognition rate of the proposed technique is high. However, the approach is still dependent on the assumption that the

(k)   Match score = 0.0000   (l)   Match score = 0.0000

(m)   Match score = 0.0000   (n)   Match score = 0.0000

(o)   Match score = 0.0000   (p)   Match score = 0.0000

(q)   Match score = 0.0000   (r)   Match score = 0.0000

(s)   Match score = 0.0000   (t)   Match score = 0.0000

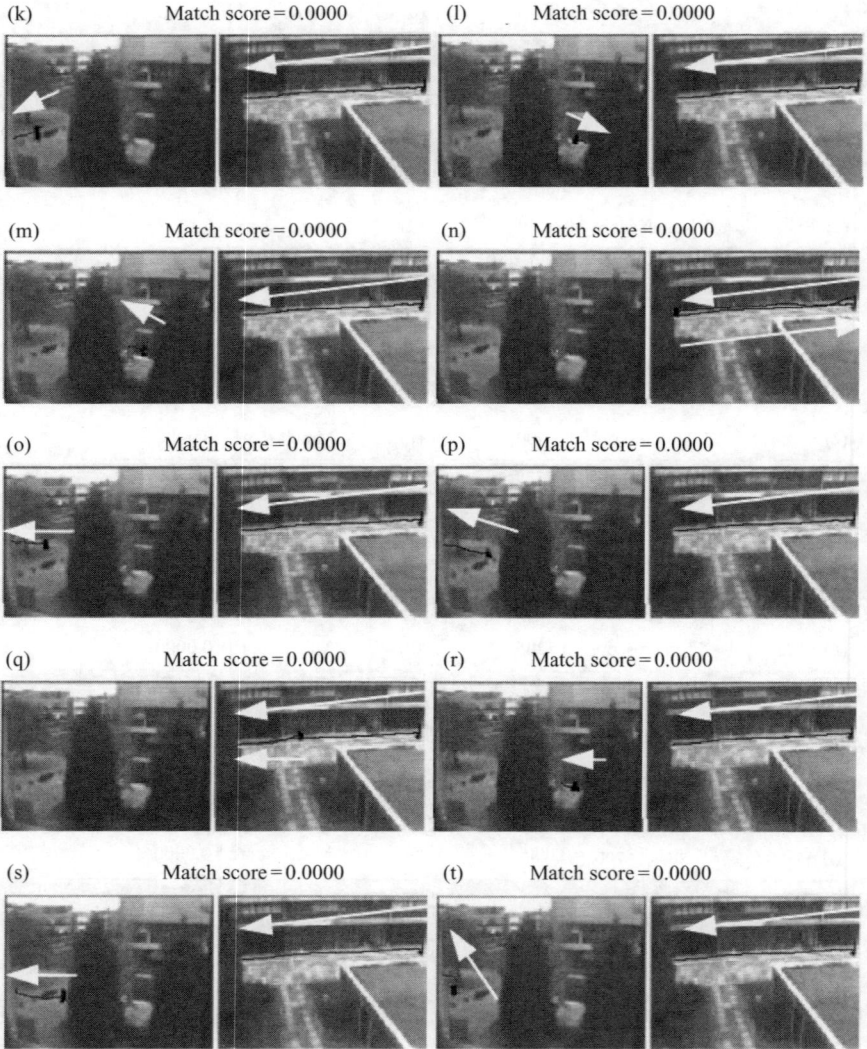

Figure 6.13b   Continued

separated regions in the same or different views are close to each other. Future work will investigate the correlation between accuracy and distance for this technique. Although we demonstrate the approach here using two cameras, it is obvious to see how the approach could be extended to larger multiple camera installations, and our future work will test this scalability.

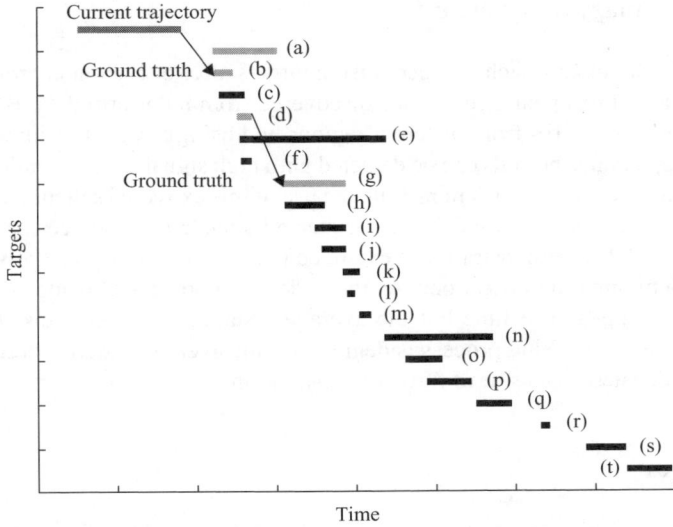

Figure 6.14   Timeline of an example of recognising reappearing target between different regions in spatially separated cameras

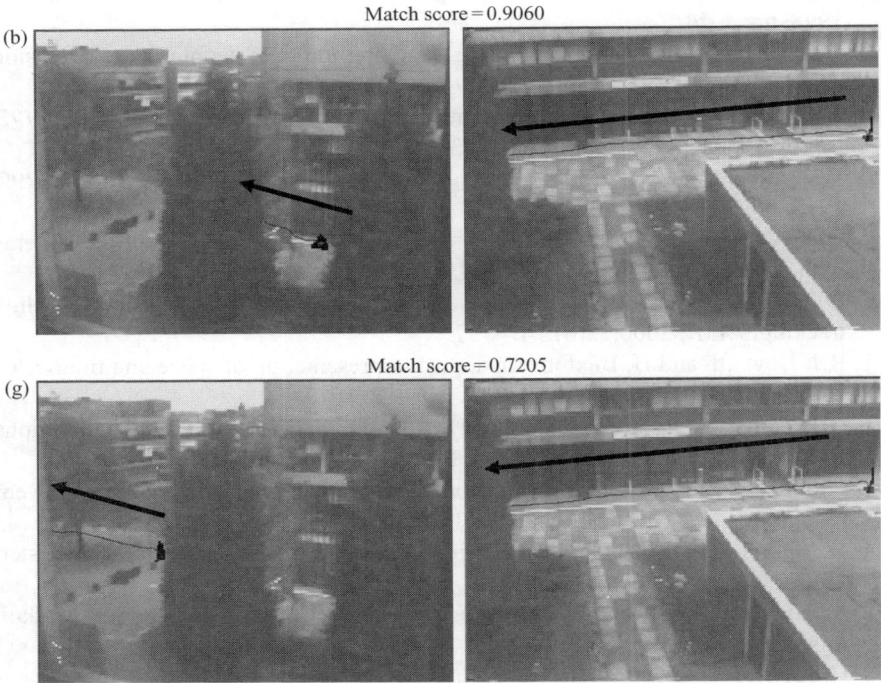

Figure 6.15   Two most likely matches between trajectories

6.10   Summary and conclusions

In this chapter, an approach for recognising targets after occlusion is proposed. It is based on salient reappearance periods discovered from long-term data. By detecting and relating main paths from different regions and using a robust estimate of noise, salient reappearance periods can be detected with high signal-to-noise ratios. Off-line recognition is performed to demonstrate the use of this extracted salient reappearance period and the appearance model to associate and track targets between spatially separated regions. The demonstration is extended to regions between spatially separated views with minimal modifications. As the underlying process of reappearance is not the salient reappearance time but the average distance between paths, the performance of this recognising process is degraded if the average distance between paths is increased. These issues need further investigation.

References

1 T. Kanade, R. Collins, A. Lipton, P. Anandan, P. Burt, and L. Wixson. Cooperative multi-sensor video surveillance. In *DARPA Image Understanding Workshop*, 1997, pp. 3–10.

2 T. Kanade, R. Collins, A. Lipton, P. Burt, and L. Wixson. Advances in cooperative multi-sensor video surveillance. In *DARPA Image Understanding Workshop*, 1998, pp. 3–24.

3 T.H. Chang, S. Gong, and E.J. Ong. Tracking multiple people under occlusion using multiple cameras. In *BMVC00*, 2000, Vol. 2, pp. 566–575.

4 T. Huang and S. Russell. Object identification in a Bayesian context. In *IJCAI97*, Nagoya, Japan, 1997, pp. 1276–1283.

5 V. Kettnaker and R. Zabih. Bayesian multi-camera surveillance. In *CVPR99*, 1999, pp. 253–259.

6 O. Javed, Z. Rasheed, K. Shafique, and M. Shah. Tracking across multiple cameras with disjoint views. In *ICCV03*, 2003, Vol. 2, pp. 952–957.

7 C. Stauffer and W.E.L. Grimson. Learning patterns of activity using real-time tracking. *PAMI*, 2000;22(8):747–757.

8 R.J. Howarth and H. Buxton. Analogical representation of space and time. *IVC*, 1992;10(7):467–478.

9 J.H. Fernyhough, A.G. Cohn, and D.C. Hogg. Generation of semantic regions from image sequences. In *ECCV96*, 1996, Vol. 2, pp. 475–484.

10 N. Johnson and D. Hogg. Learning the distribution of object trajectories for event recognition. *IVC*, 1996;14(8):609–615.

11 D. Makris and T. Ellis. Finding paths in video sequences. In *BMVC01*, Manchester, UK, 2001, pp. 263–272.

12 V. Nair and J.J. Clark. Automated visual surveillance using hidden Markov models. In *VI02*, 2002, p. 88.

13 C. Stauffer. Automatic hierarchical classification using time-based co-occurrences. In *CVPR99*, 1999, Vol. 2, pp. 333–339.

14 J. Shi and J. Malik. Normalized cuts and image segmentation. *PAMI*, 2000;22(8):888–905.

15 P. KaewTraKulPong and R. Bowden. An improved adaptive background mixture model for real-time tracking with shadow detection. In Video-based Surveillance Systems: Computer Vision and Distributed Systems. P. Remagnino, G.A. Jones, N. Paragios, and C.S. Regazzoni, Eds. Springer: Berlin, 2002, pp. 135–144, ISBN 0-7923-7632-3.

16 P. KaewTraKulPong and R. Bowden. An adaptive visual system for tracking low resolution colour targets. In *Proceedings of British Machine Vision Conference, BMVC01*, 2001, pp. 243–252.

17 P. KaewTraKulPong and R. Bowden. A real-time adaptive visual surveillance system for tracking low resolution colour targets in dynamically changing scenes. *IVC*, 2003;21(10):913–929.

18 T. Horprasert, D. Harwood, and L.S. Davis. A statistical approach for real-time robust background subtraction and shadow detection. In *Frame-Rate99 Workshop*, 1999. http://www.vast.uccs.edu/~tboult/FRAME/Horprasert/HorprasertFRAME99.pdf.

19 B. Berlin and P. Kay. *Basic Color Terms: Their Universality and Evolution*. University of California, 1991.

20 J. Sturges and T.W.A. Whitfield. Locating basic colours in the Munsell space. *Color Research and Application*, 1995;20(6):364–376.

21 J. Black, T.J. Ellis, and D. Makris. Wide area surveillance with a multi-camera network. In *IDSS-04 Intelligent Distributed Surveillance Systems*, 2003, pp. 21–25.

22 P. KaewTraKulPong. *Adaptive Probabilistic Models for Learning Semantic Patterns*, PhD Thesis, Brunel University, 2002.

23 S. Blackman and R. Popoli. *Design and Analysis of Modern Tracking Systems*. Artech House, 1999.

Chapter 7

A distributed multi-sensor surveillance system for public transport applications

J-L. Bruyelle, L. Khoudour, D. Aubert, T. Leclercq and A. Flancquart

7.1 Introduction

Public transport operators are facing an increasing demand in efficiency and security from the general public as well as from governments. An important part of the efforts deployed to meet these demands is the ever-increasing use of video surveillance cameras throughout the network, in order to monitor the flow of passengers, enable the staff to be informed of possible congestion and detect incidents without delay. A major inconvenience of this approach, however, is the very large number of cameras required to effectively monitor even a comparatively small network. Added to the cost of the cameras themselves, the cost and complexity of the required wiring, plus the sheer impossibility of watching all the images at the same time, make such a system growingly ineffective as the number of cameras increases.

In recent years, image-processing solutions have been found to automatically detect incidents and make measurements on the video images from the cameras, relieving the staff in the control room of much of the hassle of finding out where interesting events are happening. However, the need remains to bring the latter to a centralised computer located in a technical room, which leaves the need to place huge lengths of video cabling, increasing the cost and the complexity and decreasing the adaptability of the video system.

In the framework of the EU's PRISMATICA programme [1], INRETS devised and tested in real-life conditions an architecture to address these problems. The general idea is to avoid sending many full-resolution, real-time images at the same time to the video processor, by delegating the processing power close to the cameras themselves, and sending only the meaningful images through the general network to the control room. Until recently, computers and video grabbers were much too expensive to even dream of having multiple computers spread all over the network. But costs are decreasing at a steady pace, and it is becoming realistic to believe that such a thing will be commonplace soon. Existing technologies already allow, although still at a cost, realising such a working network.

7.1.1   General architecture of the system

Unlike similar older-generation systems using a centralised computer linked to all the cameras via coaxial cables, the presented solution takes advantage of the ever-decreasing cost and size of the industrial computers, and uses several local processors each linked to a small number of cameras (ideally one for each processor). Each processor is linked, via a general-purpose bus (Ethernet in our case), to a supervising computer located in the control room (Figure 7.1). This architecture offers several advantages over the older ones:

- Due to the small number of cameras connected to each frame grabber, the need to choose between processing one camera quickly or processing several cameras

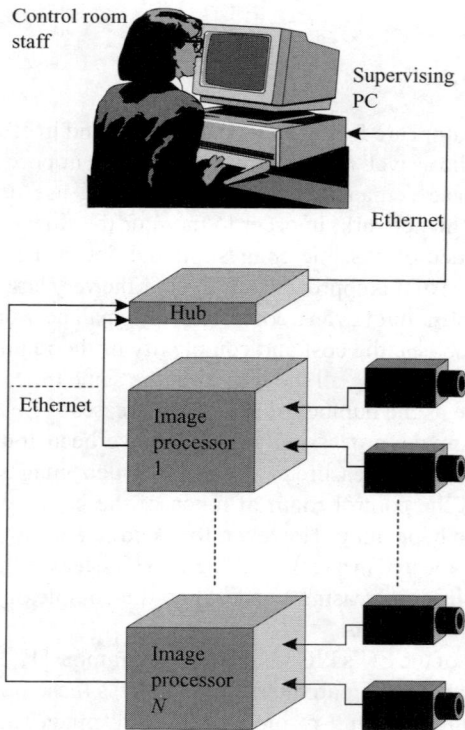

Figure 7.1   General architecture of the camera network

slowly is drastically reduced, even eliminated, if the operator can afford one processor for each camera (which will become cheaper and cheaper with time).

- Processing locally allows sending data over the cables only when needed – that is, when alarms occur – instead of sending the full video stream all the time. This decreases considerably the required bandwidth and allows using standard computer networks, such as Ethernet, which substantially reduces the cost and complexity of the cabling.
- Allowing the use of a distributed network, instead of the usual point to point connection scheme, makes it possible to just use the closest standard Ethernet sockets where needed, and make changes – for example, move a camera to another location – without having to rewire the video network.

7.2   Applications

Several functions have been implemented in a local camera network, which can be divided into two classes: incident detection functions and passenger flow measurement functions.

7.2.1   Incident detection functions

These functions detect abnormal situations, such as intrusions in forbidden areas or left luggage, and then send a message to the control room, along with details for the staff to evaluate the situation, namely the image that has raised an alarm, and a highlighting of the part of the image where the incident lies.

The functions that have been implemented are intrusions in forbidden or dangerous areas and abnormal stationarity.

7.2.2   Passenger flow measurement functions

The purpose of these functions is to gather statistics related to the traffic of passengers, both as a means of easing the mid-term and long-term management of the public transport operators (the data are added to a database which can be exploited using software such as a spreadsheet), and as an immediate warning system for raising alarms when too many passengers are waiting, for example, at the ticket counters, so appropriate steps can be taken more effectively. These 'alarms' are, of course, of a different nature from those raised by incident detection functions, as they usually require sending more commercial staff, rather than sending security or medical staff.

The measurement functions which have been incorporated in the local camera network are counting of passengers and queue length measurement.

7.3   Intrusions into forbidden or dangerous areas

The aim of this application is to improve safety by detecting, automatically and in real time, people entering areas in which they are not allowed or that are hazardous.

This type of incident can imply malevolent behaviour, but also includes events such as a fire, causing people to flee by any available exit. For this application, detection speed is crucial and the use of cameras, as opposed to conventional sensors (e.g. door switches or optical barriers), allows more flexibility (one camera can cover more surface than a door switch, and can be reconfigured by software much more easily than an optical barrier which requires physical reinstallation) and can provide more information to the security staff, by providing them with images of the incident.

For our purpose, intrusion is visually defined as an 'object' (in the most general sense of the term), having the size of a human being, moving in the forbidden zone.

This definition requires solving three separate problems:

- Detecting something that moves.
- Measuring its size.
- Deciding whether or not it is in the forbidden area.

All these need to be done in real time, with the available processing power (which is not much in the case of our miniature local processors) and with the objective of no missed alarms and as few false alarms as possible.

7.3.1   Camera set-up – defining the covered area

Unlike conventional video surveillance systems where one only wants to see the area to cover, the way to place and aim the camera is a critical part of designing the image-processing based system, as computers work in a much less sophisticated way than the human visual cortex.

Providing an unambiguous image, in particular free from overlapping or occluding subjects and perspective effects, is particularly important for achieving correct detection and reducing the rate of false alarms without sophisticated (and lengthy) pre-processing. In a general way, the camera is best placed to detect intrusions when, looking at the video picture, one can say something along the lines of 'this part of the image *is* the forbidden area, so I can say with certainty that anything entering this part of the image *is* an intruder'.

In the present case (Figure 7.2), the forbidden area is a door to the tarmac of Newcastle International Airport, which obviously should not be walked through except for service or emergency reasons. It is, on the other hand, safe to walk in front of it, and this should not raise a false alarm.

The camera is placed just above the door, aimed vertically so that the lower part of the image is the area under surveillance, while the upper part does not raise an alarm (Figure 7.3), so people standing as close as a few centimetres from the door are not mistakenly seen in the forbidden area and thus cannot raise a false alarm.

7.3.2   Extracting the moving objects

A specific algorithm is used to perform the detection of moving edges. This algorithm (nicknamed STREAM, a French acronym for 'real-time motion extraction and analysis system') has been previously designed by INRETS/USTL for surveillance

Figure 7.2   The test site – a door to the tarmac of Newcastle Airport. The camera is in the missing part of the ceiling

Figure 7.3   The view provided by the camera

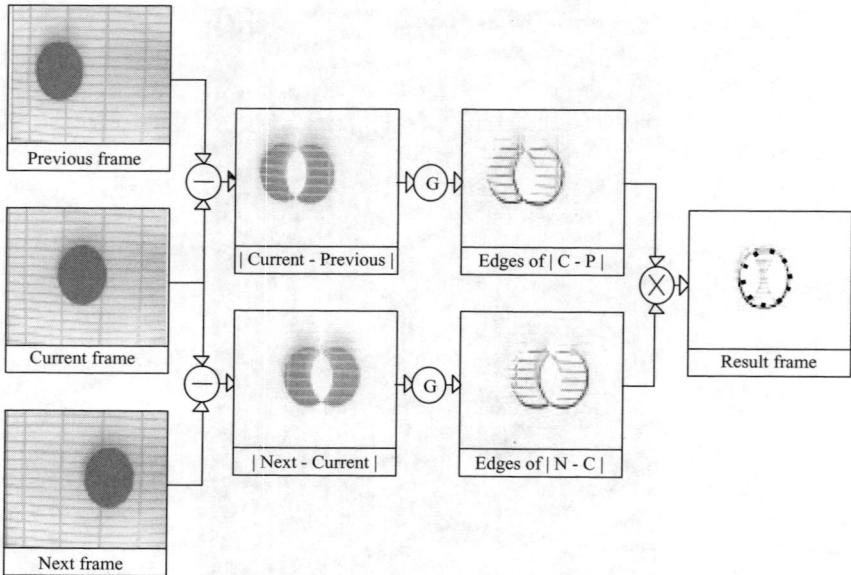

Figure 7.4   Moving edge detector algorithm (STREAM)

of cross-roads [2]. Thus, it deals especially well with real-life conditions (uncontrolled lighting changes of background, unknown moving objects). This algorithm is based on the analysis of the differences of grey level between successive frames in an original video sequence of images (Figure 7.4).

Edge extraction, performed on the difference of successive images, allows it to retain only the edges of moving objects in the frame. A refinement makes use of three successive images to avoid artefacts typical of other similar methods.

Figure 7.5 shows in its left part an image shot by a camera in real-life conditions, whereas the right part shows the moving edges found in the same image. The moving edges are available as a binary image containing white pixels where a moving edge is found. All the other pixels are black.

The local computer runs the STREAM algorithm. Interestingly, although the STREAM was designed to be easily implemented in hardware, and the associated processor was indeed used in precedent applications using the classical, centralised processing scheme, this was proven unnecessary in the new scheme in which only one or two cameras are connected to the local image processing computer, and all the processing is done in software, thus reducing the implementation cost.

7.3.3   Defining the size of objects

The pictures provided by the STREAM show very clearly the edges of anything (or anyone) moving in the field of the camera. Small objects, for example, bus tickets thrown on the floor, are easily detected by the STREAM and must not be recognised

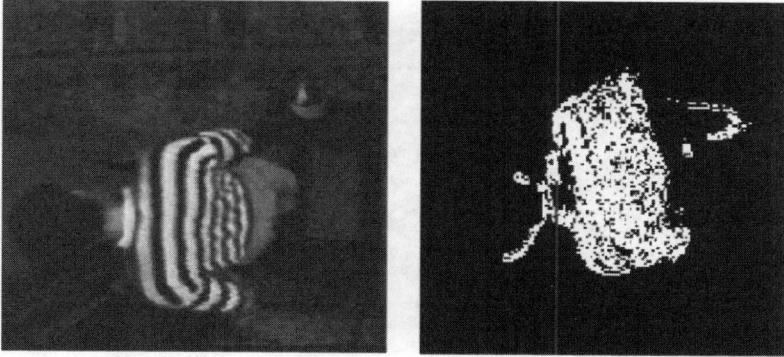

Figure 7.5   Example of moving edge detection

as incidents. The next step for good robustness is therefore to determine the size of the moving objects [3]. It has been agreed with operators that an object smaller than a small dog should be discarded.

To perform this task, we extract all the contours formed by the moving edges provided by the STREAM algorithm, using a suitable contour-following algorithm. Then the surrounding rectangles are calculated for each contour, and those that overlap are merged to build the global objects. This allows taking into account several objects simultaneously present in the image.

7.3.4   Forbidden area

The forbidden areas (several of these can be defined in the image) are defined by the user as a set of one or more rectangles.

This type of primitive shape was chosen for the fast computation it allows. For instance, the test to check whether a point belongs to a rectangle consists of just four comparisons:

If (X <= Xright)
    And (X >= Xleft)
    And (Y <= Ytop)
    And (Y >= Ybottom)
        Then [the point is in the rectangle]

No other shape offers this simplicity.

Preliminary tests showed that, for practical purposes, any shape can be represented by a few rectangles (Figure 7.6). Plus, it is an interesting shape regarding the human–machine interface (HMI), as it is easy for the operator to define the forbidden area on-screen with a few mouse clicks.

In order to check whether an intrusion is happening, we test for inclusion of the surrounding rectangles of the moving objects in the forbidden area. If inclusion does exist, then there is intrusion, and an alarm is raised.

Figure 7.6   Definition of the forbidden areas using rectangular shapes

7.3.5   Usage of the network

As was said above, all the image processing is done locally, yielding minimal load of the network. But of course the network is still required in two cases:

- when an alarm is raised (information from the local processors to the supervising computer);
- to set the processing parameters (parameters from the supervising computer to the local processors).

When an alarm is raised, the local processor tells the supervising computer that an incident is in progress, a minimal requirement for an incident detection system. It can also (on request only, to avoid using up unnecessary bandwidth) send the still image that triggered the alarm, in order to give more information to the operators in the control room. Moreover, the local processor records continuously images on its local hard disk, in what can be thought of as a FIFO-like video recorder, and can send a video sequence covering a few seconds before and after the moment when the incident was detected. The supervisor can download these still images and video sequences onto his or her own hard disk to document the incidents. The supervising computer runs the appropriate HMI to manage the way the alarms and the associated information are presented to the operators.

It is also possible, on request from the supervising computer, to download images from the camera even when no alarm is raised, for routine monitoring purpose. However, it is hardly conceivable to feed actual 'screen walls' by this means, as the available bandwidth on the Ethernet would be too low (or the image compression too drastic).

The local processor needs several parameters (the minimum size of objects that can trigger an alarm, the forbidden area, etc.). These parameters are best determined in the control room, by a few mouse clicks over an image sent by the camera via the

Figure 7.7   Intrusion image – no intrusion

Figure 7.8   The same image processed – some noise, but no detection

local processor and the network. Then the supervising computer sends the parameters to the local.

7.3.6   Test results

The results shown below have been obtained on the site described above, during field trials carried out in the framework of the PRISMATICA project. In the following figures (Figures 7.7–7.16), two images are shown for each example:

- The original image, as seen by the camera.
- The processed image. The fixed areas of the scene are black, and white parts have been detected as areas affected by a movement. The area corresponding to a moving object or person is surrounded by the rectangle used to determine the size.

Figure 7.9   A roll of gaffer tape rolling in the covered area (not an intrusion)

Figure 7.10   The same image processed – the roll is detected, but too small to raise an alarm

These examples illustrate the performances of the detection system.

All sorts of objects, including newspapers, tickets, balls and handbags, have been thrown at different speeds in the area under surveillance. Such small objects were always clearly detected. As shown on Figures 7.9 and 7.10, small objects (in this example, a roll of gaffer tape, 5 × 10 cm) were accurately detected and sized using their surrounding rectangle. A simple threshold on this size has proved to be very effective in discriminating between actual intrusions and small objects.

In Figures 7.11 and 7.12, a man is passing in the field of the camera, close to the forbidden area. He is clearly detected, and he is seen as a single moving area by the

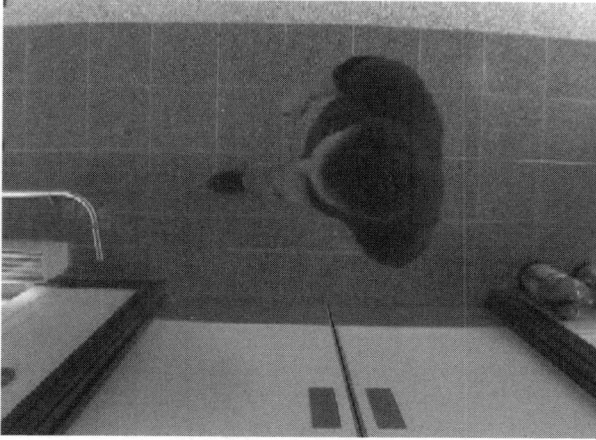

Figure 7.11   A person walking close to (but not in) the forbidden area

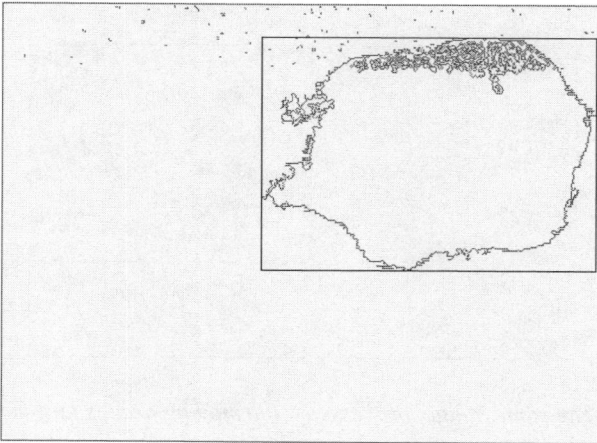

Figure 7.12   The same image, processed

system. However, he does not raise an alarm because he does not enter the forbidden area, which is the part of the image below the bottom of the door.

In order to assess the overall quality of the intrusion detection system, the latter was installed at a 'realistic' test site at Newcastle International Airport. The test was carried out in two stages, intended to measure the false alarm rate and then measure the rate of missed alarms.

The rate of false alarms was measured by allowing the equipment to run on its own for a duration of time, in operational conditions, during which (as expected) no passengers came beneath the camera to cause false detection. In order to 'tickle' the system a little more, we then sent a member of the team to walk close to the edges

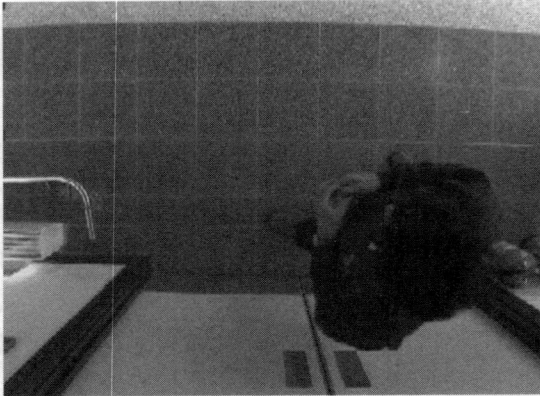

Figure 7.13   An intruder, walking in front of the door

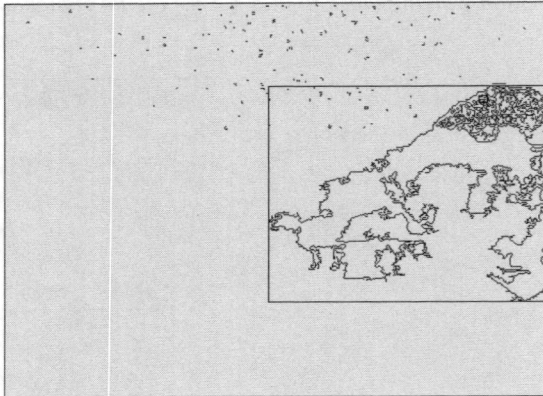

Figure 7.14   The same image, processed – the moving area is large enough to raise an alarm

of the covered area (Figure 7.11), in order to cast shadows or reflections to the floor, which might trigger some false alarms.

Also, small objects were thrown on the floor (Figure 7.9) to adjust the size threshold of the detection algorithm to distinguish between people (who should raise an alarm) and small objects (which should not). These sequences were also used as test sequences to estimate the occurrence of false alarms caused by small objects.

Finally, the rate of missed alarms was measured by sending members of the team to walk, run and stop beneath the camera, in the covered area. The rate of missed alarms was defined as the proportion of people who entered and left the frame without raising an alarm.

Figure 7.15 An intruder, standing motionless in front of the door. Only his hands are moving a little

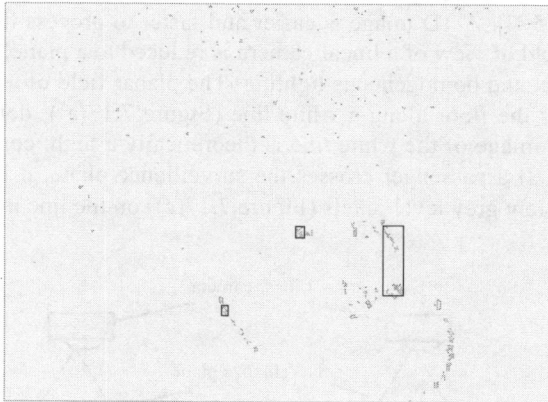

Figure 7.16 The processed image does show some moving areas, but too small to raise an alarm

The false incident rate is zero, meaning that no false detection occurred. Likewise, no missed alarm occurred over a sample of about 100 incidents: all the people who entered the covered area were detected as they entered and as they left (Figures 7.13 and 7.14). In some tests, during which they were asked to stand still in front of the door, they remained undetected for a few video frames (Figures 7.15 and 7.16) but were detected again as soon as they resumed moving.

7.3.7 Conclusion

The results of the trials show that the output of the algorithms implemented on the local camera network meets the expectations, in particular with a good robustness to lighting conditions (e.g. alternating sun/cloud) and to the processing parameters, which do not seem to need adjustments during normal operation.

7.4   Counting of passengers

The considerable development of passenger traffic in transport systems has quickly made it indispensable to set up specific methods of organisation and management. For this reason, companies are very much concerned with counting passengers travelling on their transport systems [4]. A passenger counting system is very important especially on the following points: best diagnosis on the characteristics of fraud, optimisation of lines management, traffic control and forecast, budgetary distribution between the different lines, improvements of the quality of service.

7.4.1   Counting system overview

In many applications of computer vision, linear cameras may be more adapted than matrix cameras [5–7]. A 1D image is easier and faster to process than a 2D image. Moreover, the field of view of a linear camera is reduced to a plane, so it is easier to maintain a correct and homogeneous lighting. The planar field of view of the linear camera intersects the floor along a white line (Figure 7.17(a)), defining a surveillance plane. The image of the white line is theoretically a high, constant grey level (Figure 7.17(b)). If a passenger crosses the surveillance plane, it leaves a shadow characterised by low grey level pixels (Figure 7.17(c)) on the line image.

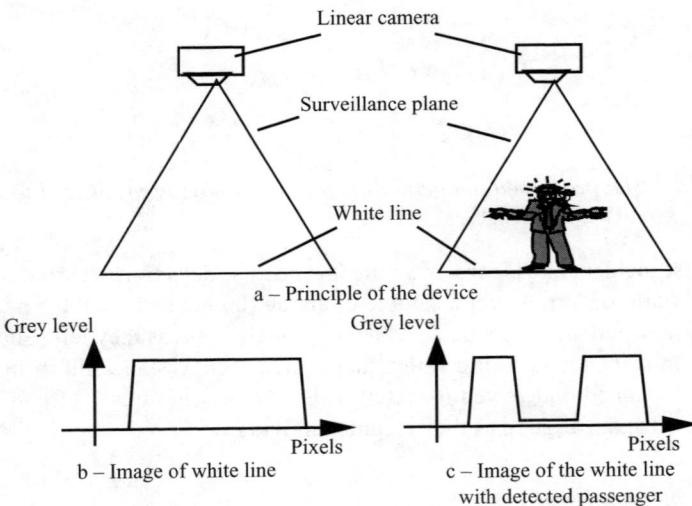

Figure 7.17   Principle of the system

| The counting camera (right) with its infrared lighting system (221 LEDs). The camera on the left is used to record images for reference manual counting. | The counting camera installed at the test (Newcastle International Airport). The two thin white stripes on the floor are the retro-reflective lines. |

Figure 7.18   The imaging system used by the counting system

In real-world conditions, however, under daylight or artificial light, and due to the reflecting characteristics of any passenger crossing the surveillance plane, the signal is not as simple as the theoretical one. The efficiency of the detection is improved by ensuring that the passenger as seen by the camera is darker than the background. This is achieved by:

- using retro-reflecting material instead of the white line, and a lighting system next to the lens, so the light that hits the retro-reflecting line is entirely reflected back to the lens; and
- adjusting the illumination level according to the reflecting material characteristics, using infrared diodes and appropriate filters on the lens, so only the light emitted by the system is seen by the camera.

The final setup is shown in Figure 7.18.

7.4.2   Principle of the counting system

This active camera yields a low response level, even if a very bright object crosses the plane under high lighting environmental conditions. Therefore, the objects of interest have characteristic grey levels significantly different from those of the background, so mere grey-scale thresholding provides reliable segmentation. The result of this segmentation is a binary line image [8]. Figure 7.19 shows the space – time image sequence recorded by the camera, where successive binary line images are piled up. The space is represented by the horizontal axis and the time is represented by the vertical one.

The width of the shape of a pedestrian along the horizontal axis is the image of his or her actual size. In the vertical direction however, the depth of the shape depends on the speed of the pedestrian. This sequence of piled line images gives a shape that

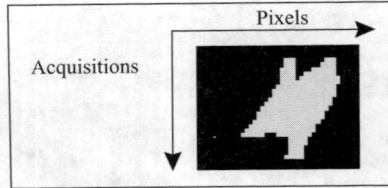

Figure 7.19   Line image sequence

represents the motion, the position and the width of a pedestrian crossing the plane. If he or she walks slowly, he or she will be seen a long time by the camera and the corresponding shape will be very long, that is, made of many successive line images. On the other hand, if they walk quickly, they will be seen during a small number of acquisitions, and so the shape is shorter, that is, a small number of line images.

An efficient detection of any pedestrian crossing a single surveillance plane is not sufficient for an accurate analysis of his motion. That is why we propose to use two parallel planes, so it becomes possible to evaluate the speed, including direction and module, of any pedestrian crossing the two planes. Indeed, by measuring the time lag between the appearance of a pedestrian in each of the two planes, it is easy to measure his or her speed. Finally, the order of crossing the two planes indicates the direction of the movement.

The ability of detecting people crossing the surveillance planes, matched with the knowledge of their speed provided by the two planes of the device, are the main features of our counting system. For this application, retro-reflecting stripes are stuck on the floor whereas cameras are clamped under the ceiling, so the surveillance planes are vertical and perpendicular to the main direction of the passenger flow.

7.4.3   Passengers counting

We have two types of information available to carry out the counting:

• The space information along the horizontal axis.
• The speed information which is computed at each pixel by taking into account the time lag between the crossing of the two planes at corresponding pixels, and the distance between these two planes.

7.4.4   Line images sequence of pedestrians

The two cameras yield two sequences of line images, as shown in Figure 7.20 where three pedestrians are walking in the same direction. One can note that each pedestrian appears in the two elementary sequences. Due to body deformations, the two shapes corresponding to the same pedestrian are not strictly identical. Furthermore, the shapes in the sequence associated to camera 1 appear before those seen by camera 2. This time lag is due to the time necessary to cross the gap between the two parallel

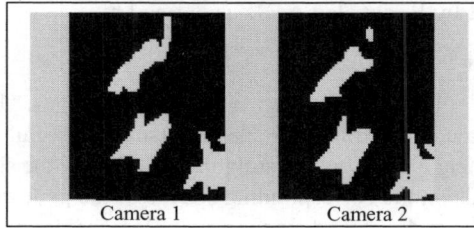

Figure 7.20   Off-peak period sequence

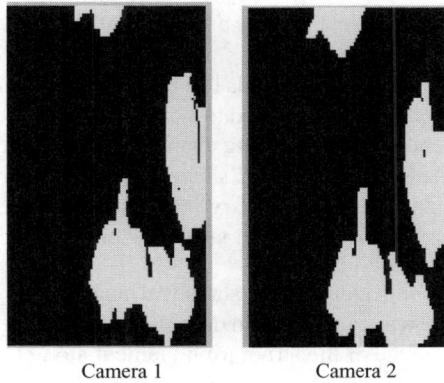

Figure 7.21   Crowded period sequence

planes. This sequence has been recorded during an off-peak period so that one shape corresponds to one pedestrian. When a shape corresponds to the image of only one pedestrian, we call it a 'pattern'.

During crowded periods, the density of pedestrians crossing the surveillance planes leads to images where the corresponding patterns are merged to give larger shapes (Figure 7.21). In this case, it is not easy to evaluate the right correspondence between shapes and pedestrians since we do not know if the shapes correspond to one, two or more patterns.

As seen in Figure 7.21, some of the shapes obtained from the two cameras correspond to more than one pedestrian.

Thus, we have developed an image processing method using binary mathematical morphology, completed and improved, to split the shapes into individual patterns, each one corresponding to one pedestrian.

7.4.5   Algorithm using structuring elements of varying size and shape

Morphological operators work with the original image to be analysed and a structuring element [9,10]. Let A be the set of binary pixels of the sequence and B the structuring element.

The erosion of A by B, denoted A⊖B, is defined by:

$$A \ominus B = \{p/B + p \subseteq A\}.$$

Usually, the morphological operations are space-invariant ones. The erosion operation is carried out using a unique structuring element (square, circular, etc.).

A single structuring element is not suitable for splitting the shapes since the area of the shapes depends on the corresponding speeds of the pedestrians.

This is why we have developed a new morphological, entirely dynamic method, dealing with the diversity of the areas of the shapes. Instead of using a single structuring element, as used in the literature, we have chosen to use a set of structuring elements to cope with the large range of speeds of the pedestrians. Since the structuring elements are different in size, the basic properties of the mathematical morphology are not verified in this case.

The whole range of observable speeds has been digitised onto N levels. Once its speed is calculated, filtered and associated to a class, each pixel of the binary image is considered to be the centre of a structuring element. If this corresponding structuring element can be included in the analysed shapes, the corresponding central pixel is marked and kept in memory. A new binary image is thus built, containing only the marked pixels. The dimensions of the SE used are defined as follows:

Width of the structuring element. A statistical analysis of all the shape widths observed in our data was carried out in order to determine an average width which is used to adjust the size of the structuring element along the horizontal axis.

Direction of the structuring element. The moving directions of the pedestrians are divided in six main directions as described in Figure 7.22. For each moving direction, a particular structuring element is applied.

Depth of the structuring element. A speed level, ranging from 1 to N, is assigned to each pixel of the binary shapes in the sequence. The depth of the structuring element along the vertical axis is controlled by this speed information as shown in Figure 7.23.

The speeds in level 1 correspond to pedestrians walking quickly or running under the cameras. The corresponding structuring element is not very deep since the shapes associated to these pedestrians are very small. The speeds in level N correspond to slow pedestrians which yield a large shape in the sequence. A deep structuring element is then necessary for separating possibly connected patterns.

We can highlight here the originality of the method: the original shapes are not eroded by a single structuring element but by a set of structuring elements. This method is a novelty, compared to the existing methods in the literature. It is completely dynamic and adaptive, which allows it to provide good counting accuracy, as we will see below.

7.4.6   Implementing the system

Both economical (line-scan cameras are expensive) and practical reasons (calibration to the two retro-reflective stripes is easier and faster to do this way) led us to use one

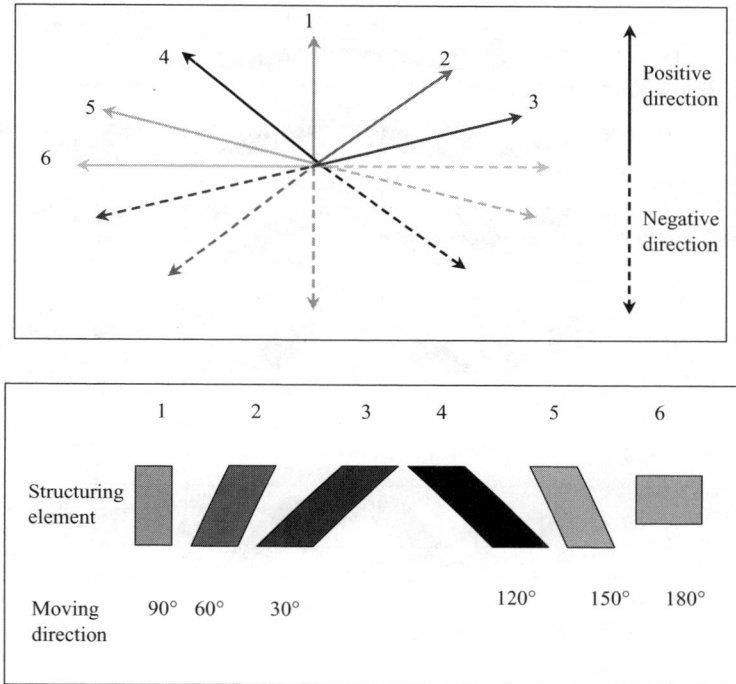

Figure 7.22   Structuring element direction

raster-scan (i.e. video) camera rather than two line-scan cameras, although all the processing is still done on two scan lines corresponding to the two retro-reflecting stripes.

However, the geometric constraints on the systems do not call for standard video cameras, because of their reduced imaging speed. Indeed, the distance between the stripes must be short enough so passengers crossing the counting system are seen simultaneously on the two stripes. Practically, this calls for a distance no longer than 10 cm or so. On the other hand, the distance must not be too short or the system will fail to measure accurately the displacement time between the two stripes and hence to determine the direction of the passenger.

The latter constraint is directly related to the frame rate of the camera. Standard video cameras have a frame rate of 25 frames per second. This means that two successive images of a passenger are 40 m apart. A passenger in a hurry, running at, say, 12 km/h,[1] covers 13 cm during the time of two successive frames, which is more than the distance separating the two frames and can prevent the system from

[1] This is a high speed for passengers in an airport, but the authors easily ran faster than that in field trials, and they are far from being professional sprinters. So the case is more realistic than it may look at first sight.

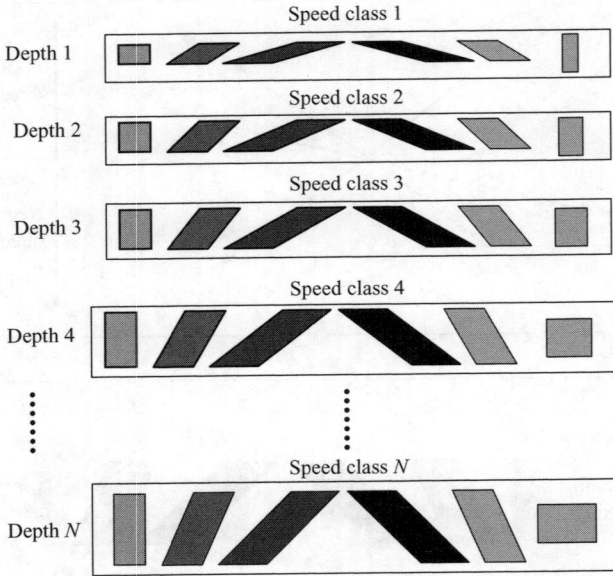

Figure 7.23   Depth of the structuring element

determining their direction. It is, however, possible to use not a standard video camera, but an industrial camera running at a higher frame rate of 100 fps. This frame rate amply solves the problem, and the cost, higher though it is, is still much lower than the combined cost of two line-scan cameras and the associated hardware.

A question that remains is that of the spatial resolution. One big advantage of line-scan cameras is their very high resolution, typically 1024 to 8192 pixels on their one scan line. Raster-scan cameras only offer less than 1000 pixels per line. However, preliminary tests over corridors in excess of 6 m wide have shown that 256 pixels instead of 1024 do not make a difference [4], while obviously offering a processing time divided by more than two.[2] So we decided to go for the 100 fps raster-scan cameras and the convenience they offer.

7.4.7   Results

7.4.7.1   Graphical results

Before giving the global results on the performance of the algorithm, it is interesting to see how the algorithm worked on some typical sequences.

In Figure 7.24, we can see typical processed results for a sequence. This figure is divided into three parts: the left and right ones represent the original binary shapes from cameras 1 and 2 and are drawn on 64 pixel-wide frames. The central part, which is represented on a 128 pixels-wide frame and in greyscale, shows the processing

---

[2] The complexity of the algorithm is $O(N \log N)$, where N is the number of pixels in the line.

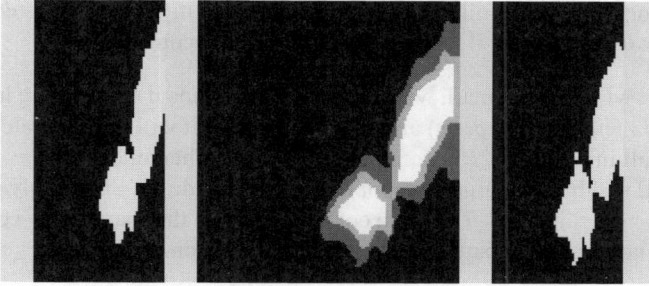

Figure 7.24   Algorithm processing

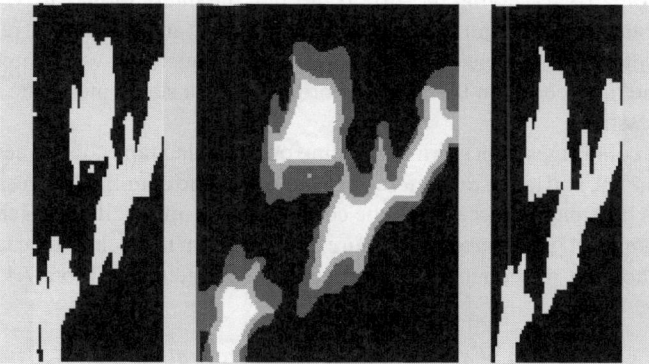

Figure 7.25   Algorithm processing during crowded condition

results – on the black background of the image, we have (brighter and brighter) the union and the intersection of the shapes from the two cameras. The white part represents the result of the erosion operation by the set of corresponding SE. Only these remaining parts are then counted. Figure 7.24 shows that the shapes previously connected have been separated by the erosion operation application. This figure shows that, since the structuring element is deformable, it maintains the geometry of the original shapes; that is why the white parts remaining have the same figure as their corresponding original shapes.

In Figure 7.25, with a higher density of shapes, one can note that the right part of the image was not separated. That is the case when the shapes are so connected that we do not know if there is one or several pedestrians. In this case, if the algorithm is not able to separate them, we will consider that there is only one pedestrian corresponding to the shapes.

Another very important fact must be mentioned here: the depth of the SE is a function of the speed detected. Concerning the structuring element width, we said above that it has been fixed as a sub-multiple of the average width. Given this rule, we must choose very carefully the width to be applied to all the shapes detected, and we must take care of the following phenomena which could happen:

- If the SE width is too small, we will be able to include it everywhere in the shapes (even in their concave parts) and the result is that it will not separate the shapes. The result in Figure 7.25 is an illustration of this phenomenon.
- If the SE width is too large, it is impossible to include the SE entirely in the initial shape. The result is the pedestrian corresponding to the shape is not counted since there is no remaining part after the erosion operation.

7.4.7.2   Statistical assessment of the system in real-world conditions

The counting system was tested at Newcastle International Airport. Images of the passengers on arrival, leaving the airport through the international arrival gate, were recorded both on the counting processor's hard disk and through a regular video camera as reference sequences. The two series of sequences were synchronised by matching the time-code (on the video recorder) and the start time of the files (on the image processor).

Manual counting was performed off-line on the video tapes, using several operators for accuracy, and taken as a reference. The automatic counting process, running in real-time on the image processor on the corresponding image files, was then adjusted to give the smallest global error (the figure on the bottom right end of the table below). Then the whole set of sequences was processed again to fill in Table 7.1.

Table 7.1   Comparison of manual and automatic counting

Seq. #	Duration (h:mm:ss:ff)	Manual count			Automatic count			Error (%)
		From planes	To planes	Total	From planes	To planes	Total planes	
1	1:21:19:01	201	3	204	200	2	202	−0.98
2	0:43:28:04	216	1	217	188	9	197	−9.22
3	0:06:12:10	25	7	32	21	8	29	−9.38
4	0:49:08:19	400	3	403	358	22	380	−5.71
5	0:21:09:18	54	0	54	52	1	53	−1.85
6	0:45:48:23	139	0	139	124	4	128	−7.91
7	0:45:39:22	191	1	192	184	2	186	−3.13
8	0:15:49:11	64	4	68	64	3	67	−1.47
9	0:20:37:21	157	4	161	139	9	148	−8.07
10	0:48:13:13	576	7	583	520	26	546	−6.35
11	0:05:21:23	4	0	4	4	0	4	0.00
12	0:23:19:07	86	0	86	82	3	85	−1.16
13	0:23:19:07	397	2	399	354	19	373	−6.52
Total				**2542**			**2398**	**−5.66**

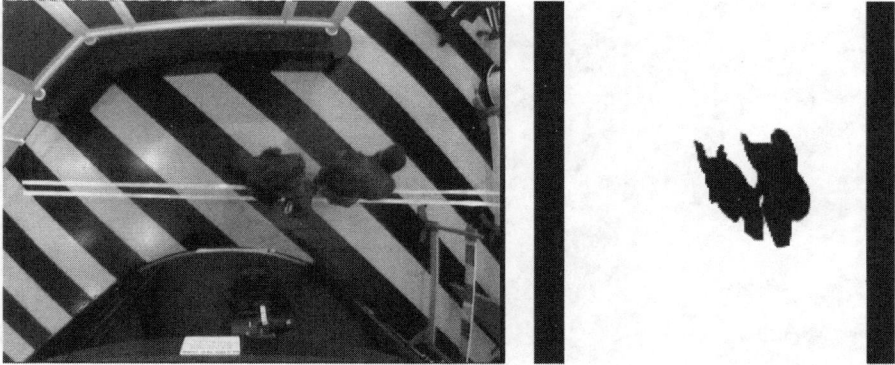

Figure 7.26　*Two passengers very close to each other can be 'merged' (right image = reflection on the lower retro-reflective stripe) and counted as one*

All the recorded sequences were processed using the same parameters (chosen as described above), and the resulting counts were compared to the results of manual counting taken as a reference. The resulting global error, over a sample of about 2500 passengers, was an underestimation by 5.66 per cent. Interestingly, this figure did not change over a fairly wide range of the tested processing parameters, which indicates a good robustness to the parameters – in other words, the operators do not need to adjust the system all the time, which would prevent convenient operation of the counting device.

A study of individual sequences showed that the systematic under-estimation of the automatic counting is related essentially to a large number of people crossing the gate very close to each other, which may yield several passengers to be counted as a single one (Figure 7.26). Other, much less frequent causes of people not being counted are cases of two passengers pushing the same luggage trolley, or children sitting on it.

These effects are partly compensated by other issues that yield *over-counting*. The most common of these, in the framework of an airport, is the widespread use of wheeled suitcases, which have a thin handle that the system cannot see. As a result, the suitcase can be counted as a passenger due to the difficulty to distinguish between the shape of the suitcase and the shape of a 'real' person (Figure 7.27).

The same problem can arise, in a very few instances, when a luggage trolley is held at the end of the arms. Then the shapes of both the trolley and the passenger can occasionally be seen as completely unrelated, and mistakenly counted as two passengers (Figure 7.28).

Finally, we once met a regular family meeting, right on the retro-reflective stripes of the counting device. Although we could not find out the exact effect of this on the counting (our software only gives the count for the whole sequence), and such instances are too rare to cause significant errors on a large scale, we feel this shows the

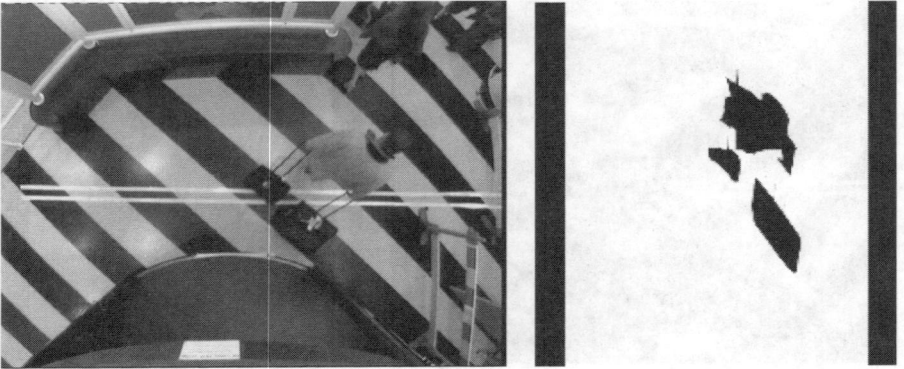

Figure 7.27   Suitcases can be seen 'detached' from the passenger (right image) and mistakenly added to the count

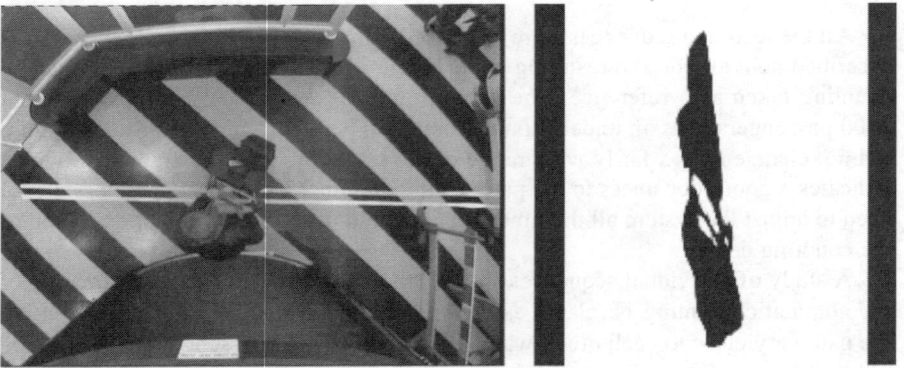

Figure 7.28   This trolley might be seen separately from the passenger and mistakenly added to the count

importance of choosing the right place to install the cameras: apparently, this one was not extremely good, and should have been chosen closer to the revolving door where such events are unlikely due to the proximity to moving parts and people coming out.

7.4.8   Implications and future developments

The accuracy found here is noticeably better than what is generally offered by existing systems (which are more in the 10–20 per cent range). Moreover, the whole system could be installed in a matter of minutes despite the need to stick retro-reflective stripes on the floor, and was shown to be remarkably insensitive to the processing and imaging parameters.

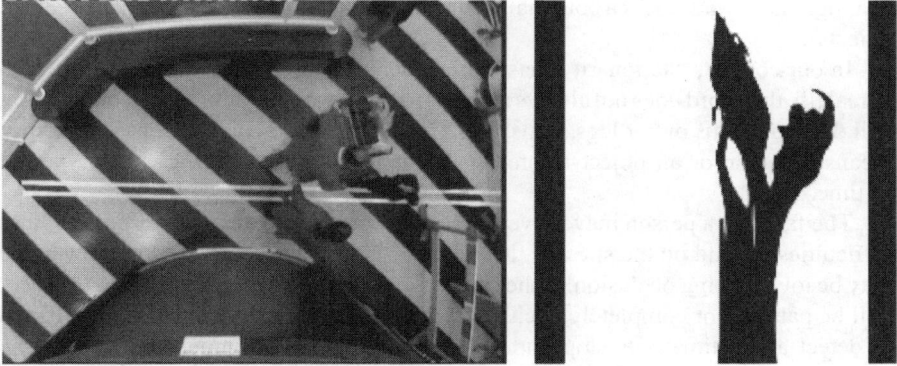

Figure 7.29   People going back and forth around the retro-reflective stripes might fool the system and cause counting errors

A major issue here is the huge quantity of parasitic light that reaches the camera in direct sunlight. The design of the system takes this issue into account, by using the following steps:

- The integrated lighting system, which provides the light to the retro-reflective system, is powerful enough to compete with direct sunlight (so white clothes lit by directs sun appear still darker than the retro-reflective stripes) and yet has enough angular coverage to cover the whole length of the retro-reflective stripe.
- Infrared LEDs (850 nm here) must be used in the lighting system, and a matched IR-pass filter (e.g. Wratten 89B) must be placed in front of the camera's lens. This allows the system to reject most of the daylight, which is the visible part of it.

Another important point is the relatively good mechanical robustness of the retro-reflective material, which was left in place from July through November without any specific protection and was not noticeably damaged by the passing of several tens of thousands of passengers (a lowest-case figure, estimated from the traffic figures provided by Newcastle Airport). Only the part of it that was lying under the brush of the revolving door was torn away by the friction. Some manufacturers offer protected material, intended to withstand such abuse, at an obviously increased cost, and with the requirement to 'bury' it into the floor, so mounting is more lengthy and expensive. However we did not have the opportunity to test such material.

7.5   Detection of abnormal stationarity

7.5.1   Introduction

This section will deal with an automatic detection of motionless people or objects using CCTV cameras. In the public transport network context, this type of detection is required to react quickly to events potentially impacting user security (e.g. drug dealing, a thief waiting for a potential victim, a person requiring assistance, unattended objects).

In our context, 'stationarity' has to be understood in the broadest sense. Indeed, in real life this word does not always mean 'completely motionless': a person standing still does move his or her legs, arms, etc. Thus, in the following sections stationarity means a person or an object located in a small image area during a given period of time.

The fact that a person may move adds complexity to the detection process. Other difficulties depend on the site and the network. For instance, the camera view angle may be low, causing occlusion. If the site is crowded, the 'stationary' person or object will be partially or completely occluded from time to time, making it more difficult to detect and estimate its stop duration. The lighting environment may affect the perception and the detection too. In fact, the method needs to be robust to contrast changes, deal with occlusion and take into account the motion of a stationary person.

In the following, after a characterisation of 'stationarity', the developed method is described. First of all, our 'change detector' is presented, with emphasis on its suitable properties. This detector is used to segment the image. Then, additional modules enable detecting excessively long stationarity. Finally, the algorithm is tested on a large number of real-life situations in subways and airports.

7.5.2   Stationarity detection system

7.5.2.1   Characterisation of stationarity

By stationarity, we imply somebody or something present in the same area for a given period of time. This definition encompasses the case of a person standing still, but moving his or her legs, arms, etc., or changing his or her position as in Figure 7.30.

A first idea to achieve the detection would be to perform shape recognition and to detect when this shape remains in the same area for a while. The benefit of the recognition process is the possibility of discriminating object classes. However,

Figure 7.30   Left – two people detected in a crowded concourse. Right – the same people after several movements and a displacement without loss of detection

the recognition is quite difficult to perform under frequent occlusions, when the view angle or the object aspect changes. This approach does not seem realistic, at least in the short term, in a crowded context.

The selected solution consists in detecting changes at the level of individual pixels and determining those 'motionless' and finally grouping together connected motionless pixels to get stationary shapes. As a drawback, the object class (person or bag, for instance) is not determined. However, the result is obtained regardless of changes in the aspect of the object and in the view angle. Another advantage is the possibility of detecting a motionless object even if only a small part of it is seen.

7.5.2.2 Change detector algorithm

The change detector constitutes a key generic piece of the overall algorithm proposed for detecting stationary objects. We assume herein that the camera is observing a fixed scene, with the presence of objects moving, disappearing and appearing. Hereafter, the term 'background' designates the fixed components of the scene, whereas the other components are referred to as 'novelties'. The purpose of the change detector is to decide whether each pixel belongs to the background or to the novelties.

In order to avoid disturbances caused by global illumination changes, one of the original ideas of our change detector is to separate geometrical information from contrast information. More specific details on the change detector algorithm can be found in Reference 11.

7.5.2.2.1 Local measures based on level lines

Any local measurement based on the level-lines geometry preserves the robustness of the image representation. We compute the direction of the half-tangent of each level line passing through each pixel. As a consequence, several orientations may be simultaneously present for a given pixel. All these orientations are then quantified for easier storage and use (Figure 7.31).

7.5.2.2.2 Representation of the images by means of level lines

To get a change detector robust to global lighting changes, we chose a contrast-independent representation of the image [10,12,13], the level line. We define

$$\aleph_\lambda I = \{P = (i,j) \text{ such that } I(P) \geq \lambda\}, \tag{7.1}$$

where $I(P)$ denotes the intensity of the image I at the pixel location P. $\aleph_\lambda$ is the set of pixels having a grey level greater than or equal to λ. We refer to the boundary of this set as the λ-level (Figure 7.32). As demonstrated in References 11 and 14, a global illumination change does not affect the geometry of the level lines but creates or removes some of them.

7.5.2.3 Reference data building and updating process

The orientations of the local level lines are used to build and update an image of the background, used as the reference. First, we count, for each pixel, the occurrence

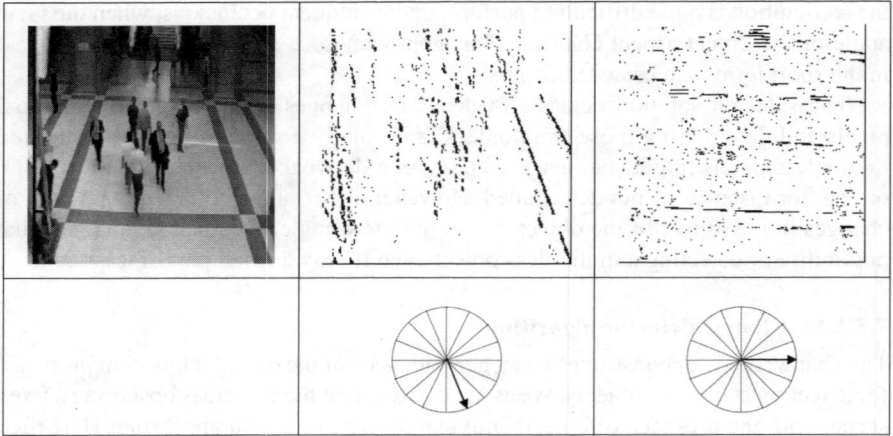

Figure 7.31   *Left – the original image. Centre and right – display of all pixels on which a level line with the quantified orientation (every $\pi/8$) passes, as indicated by the arrow inside the circle*

Figure 7.32   *Extract of the reference data. Left – one of the images used. Centre – level lines associated with the left image (only 16 level lines are displayed). Right – the pixels having at least one orientation above the occurrence threshold*

frequency of each orientation over a temporal sliding window:

$$F_{t \leq T}(P, \theta_k) = m\, F_{t \leq T-1}(P, \theta_k) + (1 - m)\, f_t(P, \theta_k), \tag{7.2}$$

with $f_t(P, \theta_k)$ a binary value indicating whether the direction θ_k exists at pixel P at time t, $F_{t \leq T}(P, \theta_k)$ the number of times the direction θ_k exists for P over a time window T, and $m = T/(T + 1)$. A direction with a large number (T_0) of occurrences is considered to belong to the reference R (Figure 7.32).

$$R_{t,T,T_0} = 1 \qquad \text{if } F_{t \leq T}(P, \theta_k) \geq T_0 \quad \text{and} \quad R_{t,T,T_0} = 0 \text{ otherwise.} \tag{7.3}$$

Figure 7.33   Left – the initial image. Centre and right – a short-term change detection without and with filtering (area opening and closing)

7.5.2.4   Change detection

Given a direction θ at a pixel P, we check if this direction occurs in the reference (R_{T,T_0}), up to its accuracy. If not, the pixel is a novelty (Figure 7.33).

$$\text{If } \exists \theta_k / f_t(P, \theta_k) = 1 \quad \text{and} \quad R(P, \theta_k) = 0 \text{ then } C(P) = 1, \text{ otherwise } C(P) = 0. \tag{7.4}$$

The time window (T) and the occurrence threshold (T_0) must be chosen by experience, depending on the context (average speed, frame rate, etc.). However, given a time window (T), a lower and an upper bound exist for (T_0) to ensure that no orientation, in the reference, is caused by noise [14]. By adjusting T, we can detect changes that occur over various time ranges. For instance, a short T will enable us to get the current motion map.

7.5.3   Stationarity detection algorithm

The extracted level-lines could be classified into one of the following categories: those that belong to the scene background and those that correspond to moving objects or to stationary objects.

The system uses the 'presence duration' to discriminate between these three categories. The background is naturally assumed to remain unchanged for a long period of time. Conversely, the moving objects will not yield stable configurations even over short periods of time. Between these two extremes, stationarity is characterised by objects that remain at approximately the same place over an intermediate period of time. This set-up then involves the use of the change detector with two time periods: a short one to detect moving objects and an intermediate one to detect moving objects and stationarities.

7.5.3.1   Detection of short-term change ('moving parts')

By applying the change detector with a short-term reference R_{short}, areas in the image containing moving objects are detected. The result is a binary image representing the short-term changes, referred to as the 'motion map' (M_t):

If $\exists \theta_k / f_t(P, \theta_k) = 1$ and $R_{\text{short}}(P, \theta_k) = 0$ then $M_t(P) = 1$

otherwise $M_t(P) = 0$.   (7.5)

7.5.3.2   Detection of long-term change ('novelties')

By using the same process with a long-term reference R_{long}, novelty areas are highlighted (N_t):

If $\exists \theta_k / f_t(P, \theta_k) = 1$ and $R_{\text{long}}(P, \theta_k) = 0$ then $N_t(P) = 1$

otherwise $N_t(P) = 0$.   (7.6)

7.5.3.3   Detection of the stationary areas

By removing the moving parts from the novelties, only the people and objects remaining stationary for at least the short-term duration are retained (S_t):

$$S_t(P) = N_t(P) - M_t(P).   (7.7)$$

7.5.3.4   Estimation of the stationary duration

Summing the occurrences of the stationary state at each pixel over time enables estimating the 'stop duration':

$\text{index\_of\_duration}(P)_t = (1 - \alpha) \times \text{index\_of\_duration}(P)_{t-1} + \alpha$ if $S_t(P) = 1$,
$\text{index\_of\_duration}(P)_t = (1 - \alpha) \times \text{index\_of\_duration}(P)_{t-1}$ otherwise.

So, for a motionless pixel P, $\text{index\_of\_duration}(P)_t = 1 - (1 - \alpha)^t, 0 \leq \text{index\_of\_duration} \leq 1$, where t is the number of previously observed images and α the value controlling the evolution speed of 'average'. α is determined such that index_of_duration is equal to a threshold T_d when the target maximal stop duration is reached. For instance, $\alpha = 1 - (1 - 0.7)^{1/(10*2*60)} = 0.001003$ considering a target stop duration of 2 min, $T_d = 70$ per cent and a 10 fps processing frame rate. T_d is chosen such that any detection is stable (index_of_duration may range from T_d to 1 without any loss of detection) and such that a detection disappears as quickly as possible when the stationary person/object leaves the scene (the detection ends when index_of_duration drops below T_d again).

In fact, to get a stable index_of_duration, its estimation is frozen when the area is occluded. In the system, the occlusion is characterised by the motion map (M_t), since it is generally caused by people passing in front of the stationary person/object. However, due to this freezing process, the estimated stop duration may be shorter than the real duration. Thus, to get the right index_of_duration, α may be increased according to the number of images for which the computation was not done ($t_{\text{occlusion}}$). Therefore, α varies in time and space:

$$t_{\text{occlusion}}(t, P) = t_{\text{occlusion}}(t - 1, P) + M_t(P)$$

$$\alpha(p)_t = 1 - (1 - T_d)^{1/t'}   \text{where } t' = t - \min(t_{\text{occlusion}}(t, P), 50\% \, t).   (7.8)$$

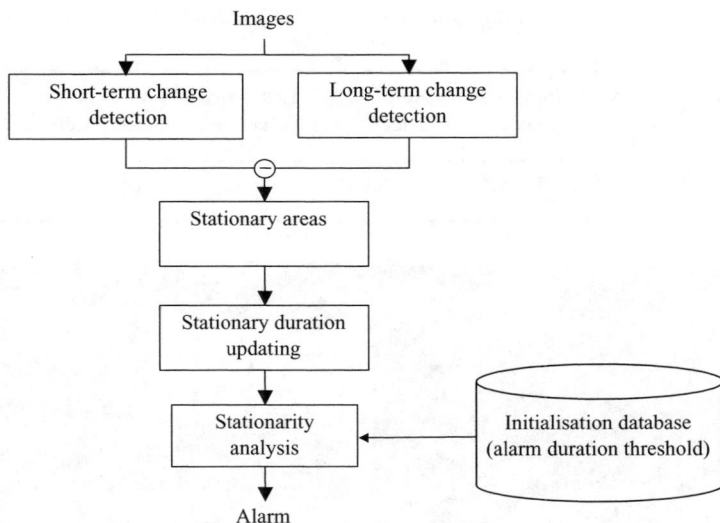

Figure 7.34   Diagram of the stationarity detection system

This definition includes a threshold on the occlusion rate to 50 per cent of the time. Indeed, for a larger occlusion duration, the number of images over which the stationarity is observed would be too low, which may yield false detection. As a consequence, there is a delay when the occlusion rate is higher than 50 per cent.

7.5.3.5   Detection of the target stationarities

Each pixel whose stationarity duration exceeds a fixed stop duration is labelled. Connected pixels are then grouped together to form a bounding box around the stationary object. Thus, the described system is able to detect stationarity regardless of the changes in the person/object position inside a small area.

Our system may be summarised by Figure 7.34.

7.5.4   Results

Our system has been tested on several real-life situations in the context of subways and airports. In order to assess the system, we manually recorded all stationarity events. Results obtained by the system were then compared with the recorded data. Table 7.2 summarises the comparison results.

As observed from these results, the system is able to deal efficiently with this detection problem (Figures 7.30 and 7.35). The non-detections can be explained by a low contrast, a stationarity duration just above the threshold or a very high occlusion rate (>90 per cent). In all but one case, the system detected the stop, but the measured stationarity duration did not reach the threshold above which an alarm is triggered.

Table 7.2   Performance of the abnormal stationarity detection system

Number of stationary situations	Number of detections	Non-detections	Erroneous detections	Detection delay (when occlusion <50%)
436	427 (98%)	9 (2%)	0	±10 s

In Paris subway   At Newcastle Airport

Figure 7.35   Examples of unattended bag detections (highlighted by white dots over the object)

7.6   Queue length measurement

7.6.1   Introduction

This section describes the automatic measurement of passenger queues using CCTV cameras. This type of research was initiated within the European CROMATICA (CROwd MAnagement with Telematic Imaging and Communication Assistance) project. It continued within the European PRISMATICA project, with emphasis on performance assessment. For this purpose the system was installed at Newcastle Airport, and several hours of video tapes were recorded during off-peak and peak periods.

Information to qualify is the estimation of both the queue length and the time spent in queue. This information is useful for at least three purposes:

- To increase the comfort of passengers by providing them with information about the average waiting time.
- To get statistics and, thus, an estimate of the crowd management efficiency. Such statistics may be used, *a posteriori*, to improve the site design and layout, to distribute the employees in a more suitable way and to raise the efficiency of ticket offices, cash dispensers, check-in desks and customs desks.
- To check continuously the effectiveness of the crowd management and react immediately to any disturbances.

Two types of queue may be encountered:

- One type corresponds to a moving flow of people, all walking in the same direction at roughly the same speed. It is the case, for instance, at stadium entrances. This type of queue may be characterised by a motion direction and the compactness of people.
- The second type is characterised by a set of people, standing generally in line (but not always), motionless for a few seconds, then walking in the same direction for a small distance, waiting again and so on until they leave the queue. It starts in specific areas such as places close to ticket offices, cash dispensers, customs and boarding gates.

The developed system deals only with the second type. It uses the measurement of stationarity time (Section 7.5) to detect when people in queue are motionless.

Detection of queues is more or less difficult depending on the environment. First of all, the lighting conditions may change over time, making it difficult to build and maintain a 'background' image. It is the case for airports, where the light comes from the outside through large windows. However, we already saw, during the presentation of the stationarity detection, that our change detector is robust to such problems. A much more difficult and specific problem is induced by the camera location: when the camera is low and not in front of the queues, a part of a queue may be occluded by a crowd and/or merged visually with another queue.

7.6.2 The queue measurement system

7.6.2.1 Characterisation of a queue

To characterise a queue in the image we postulate that people in queues are motionless at several periods and that a queue always starts at a specific location. This enables us to discriminate it from motionless people who do not belong to a queue. We will not use any hypothesis on the direction of the queue since it may evolve quickly (due to mobile fences, for instance) as we have seen during our field trials (Figure 7.36).

Figure 7.36   The direction of a queue may change drastically. Between the two situations a mobile fence has been taken away

Figure 7.37   Selected areas in the image where queues are to start

7.6.2.2   Initialisation process

To detect a queue, we need to detect regions in the image where people are motion-less for a few seconds or minutes and locate those intersecting specific locations such as ticket offices, cash dispensers, customs or boarding gates. These locations enable us to distinguish between queues (stationary regions intersecting a selected area) and isolated motionless people in the scene.

Thus, during the installation, an operator needs to tell the system where queues may start. For this purpose, we developed an HMI. The selected areas are characterised by trapezoidal shapes to take into account the perspective effect (Figure 7.37 for an example of such a selection). Each trapezium is allocated a distinctive label.

This HMI also aims at defining where the process works in the image. It allows us to reduce the computational cost and to avoid disturbances caused by queues badly perceived due to the camera location.

Since a queue is characterised by several stationarities over time, we also need to define their average duration. The latter depends on the services (customs, ticket office, etc.) and, thus, has to be set up by operators at initialisation for each type of service in their network. Thus, for each area selected, an operator indicates, through the interface, the average motionless duration (named $Stop_{duration}$ hereafter). It was set to 10 s in our context.

Finally, to get roughly the same quantity of information in the foreground and in the background of the scene, it is necessary to correct the geometric dis-tortion induced by the perspective projection. Moreover, to estimate the real queue length, it is necessary to retro-project the image of the queue to the scene. Thus, for both aspects, it is necessary to know the relationship between pixels in the image plane and points on the floor of the scene. This is done by a calibration process. Hypothesising a flat floor, it is just necessary to estimate the homography between it and the image plane. To do so, the operator selects two horizontal lines and a vertical one on the image of the floor, using the HMI, and indicates, for each of them, their real

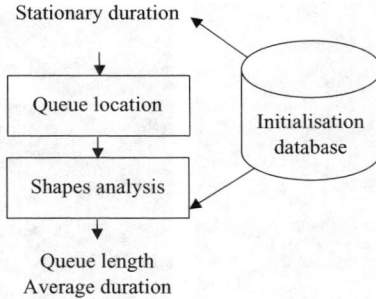

Figure 7.38   Scheme of the queue measurement system

length. They also enter the height of the camera through the HMI. Then, an algorithm automatically calibrates the system from this data.

7.6.2.3 Measurement process

Broadly speaking, the measurement is performed in two steps (Figure 7.38). Queues are extracted from their stop duration and their location.

7.6.2.3.1 Queue location

As seen previously, for each stationary pixel, the system determines its stop duration. All the connected sets of pixels classified as motionless for at least $Stop_{duration}$ are gathered to define a region REG, using the 'blob colouring' algorithm:

- $REG_t(P) = Label_{Reg}$ if the pixel P at image t belongs to the region with the $Label_{Reg}$ 'colour';
- $REG_t(P) = 0$ if it does not belong to any region.

Regions that overlap one of the selected areas (ticket office, for instance) are the queues.

To check the intersection, when the algorithm starts, an image corresponding to the selected areas is created. In this image, each pixel in a given selected area has its value set to the corresponding area label (Figure 7.39). The other pixels are set to zero (black).

The intersection computation is then a binary test:

If $REG_t(P) \neq 0$ and $AREA_t(P) \neq 0$ then queue$(REG_t(P)) = AREA_t(P)$.

At this point we may face a problem. Indeed, a region may intersect several selected areas (queue$(Label_{Reg})$ receives various area labels). In the worst case, when the view angle of the camera is low and, when the camera is not in front of the queues, two different queues may appear as one due to the perspective effect. The developed system is not able to deal with such a configuration. For this reason, the system does not consider the two queues on the right in Figure 7.40. At least, queues need to look

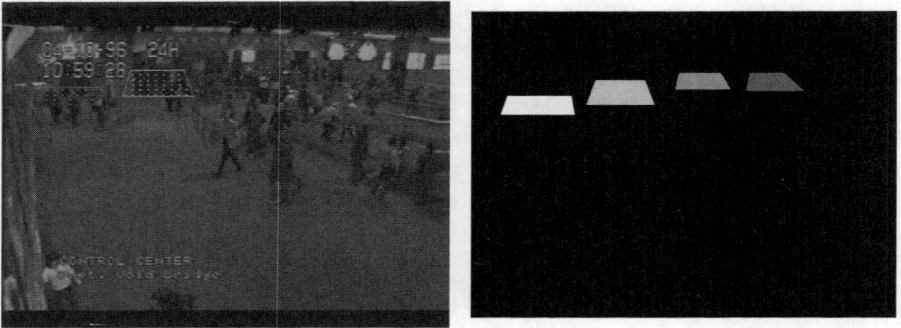

Figure 7.39   Image of the selected areas

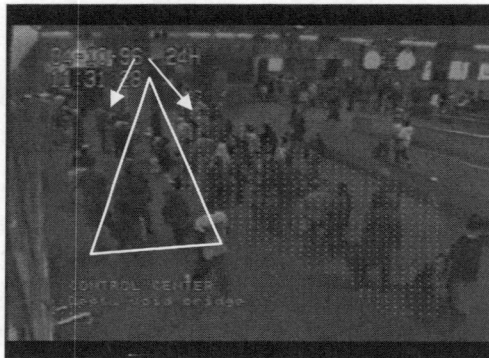

Figure 7.40   People standing between two queues. A classic region extraction process will deliver a unique region, while two queues exist

separated as in the images shot at Newcastle Airport. However, even in this situation queue regions may, from time to time, merge due to people standing between them.

The following procedure solves this problem (Figure 7.41). When a region covers several selected areas, an algorithm determines, through temporal differencing, parts (binary function A) that appear to split it into its various components.

If $\text{REG}_t(P) \neq 0$ and $\text{REG}_{t-1}(P) = 0$ then $A(P) = 1$, otherwise $A(p) = 0$.

Each new part linked with either two previously separated queues or with a queue and an 'old' (region existing for several images) stationary region is discarded.

Conversely, a queue may be split into several regions due to low contrast or a gap between people (Figure 7.42).

If these regions are close to each other, and if they were merged on the previous image, the disappearing (binary function D) parts are reinserted to get the real queue

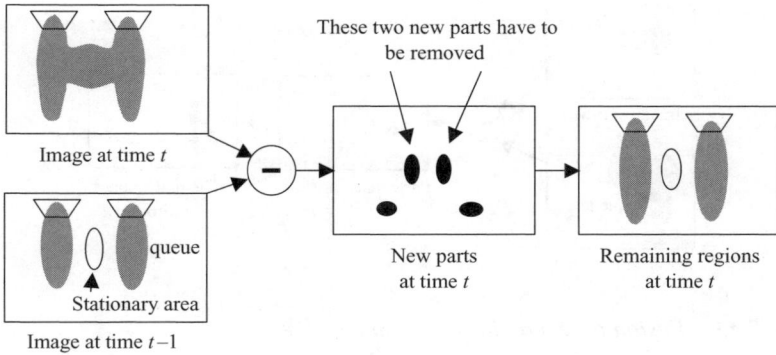

Figure 7.41   Process to split two merged queues

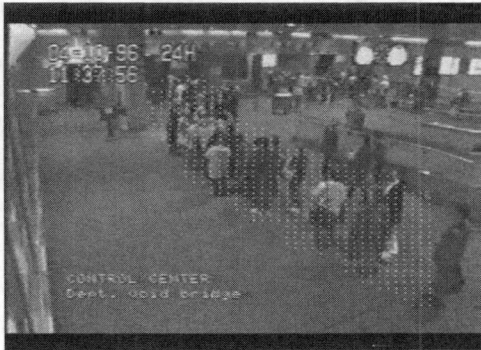

Figure 7.42   Part of a queue may be lost due to a space between people. It is not the case in our system (the queue is highlighted by white dots)

region (Figure 7.43).

If $\mathrm{REG}_t(P) = 0$ and $\mathrm{REG}_{t-1}(P) \neq 0$ then $D(P) = 1$, otherwise $D(p) = 0$.

$$(7.9)$$

7.6.2.3.2   Shape analysis

For each queue region, the main axis is extracted. Since the border of the queue region is not smooth and holes may appear inside the region due to low contrast, the axis is quite noisy. To smooth it we use both a median filter and a morphological process. Finally the axis, as well as the queue regions, are projected on the scene floor thanks to the calibration parameters to get an estimate of their real length.

7.6.3   Results

To assess the system, the difference between the real queue length and the estimated one is measured. For each tape recorded, we manually keep a cursor at the tail end of

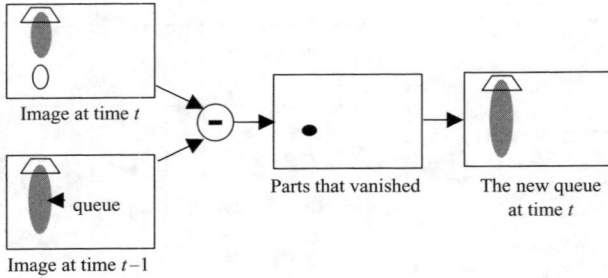

Figure 7.43   Fusion of two regions to form a queue

Figure 7.44   Examples of detected queue axes at Newcastle Airport

a queue. Every tenth of a second, its position (in pixels) is recorded. In parallel, the system determines, at the same rate, its own estimate of the position. The absolute average distance between the 'true' location versus the estimated one is then measured. An average of all distances gives the global length error.

The system was tested on more than 12 h of videos covering airport scenes during off-peak and peak hours. Thus, we obtained various types of queues, ranging from short ones to quite long ones. Figure 7.45 shows an example of a manual measurement versus the corresponding automatic one.

There are two explanations to the errors:

• There is a delay between the creation of a queue and its detection, due to the time to detect a stationarity ($Stop_{duration}$ in the worst case, off-peak period).
• The discrimination between queues and other stationary areas is not always perfect.

7.7   Conclusion

This chapter presented the distributed camera network designed by INRETS. Four applications have been implemented on this network for the purpose of

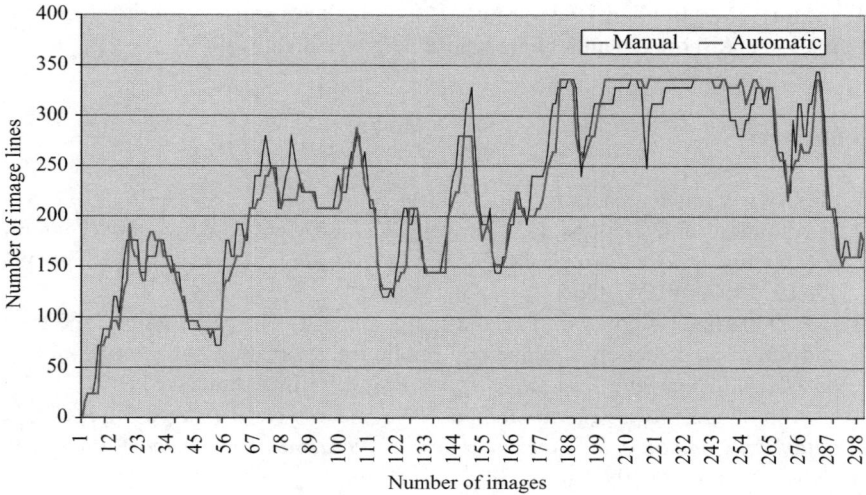

Figure 7.45 Manual queue length measurement versus automatic measurement. The average absolute error is 1.8 lines on this example (0.8 per cent considering the average length of 226 lines)

incident detection and statistical monitoring of passenger flows in public transport systems.

The whole system was tested in real-life conditions, mainly at Newcastle Airport, Heathrow Airport and Paris and Milano undergrounds. The results of the trials show that the system meets the expectations expressed by the operators, regarding ease of use, availability, effectiveness and robustness to the ambient conditions.

It must be mentioned, however, that although an HMI was implemented, it could not be properly tested, due to constraints related to the PRISMATICA project which was not aimed at industrialisation. The outputs shown here are 'technical' displays, not intended to be watched by the operation staff, but for evaluation purpose only.

As a research institute, what we propose is mainly detection algorithms. Our domain is to use cameras and computers to perform the desired detection functions. So, at the end of PRISMATICA, what we propose is a laboratory version of the demonstrators. The system is being improved in terms of the HMI and integration to the operation of a control room. For this purpose, industrial development is also under discussion with equipment manufacturers, which should yield commercial deployment in the future.

Acknowledgements

We particularly thank the public transport operators who supported us in defining, developing and testing the distributed camera network and the applications: Régie Autonome des Transports Parisiens (RATP), Azienda Trasporti

Milanesi (ATM), London Underground Ltd, British Airports Authority, Newcastle International Airport Ltd and others whose help was invaluable.

References

1 http://www.prismatica.com.
2 L. Khoudour, J.-P. Deparis, J.-L. Bruyelle *et al*. Project Cromatica. In *9th International Conference on Image Analysis and Processing*, Florence, Italy, September 17–19, 1997, pp. 757–764.
3 J.-P. Deparis, L. Khoudour, F. Cabestaing, and J.-L. Bruyelle. Fall on the tracks and intrusion detection in tunnels in public transport. In *EU-China Conference on ITS and Transport Telematics Applications*, Beijing, June 3–6, 1997.
4 L. Khoudour. *Analyse Spatio-Temporelle de Séquences d'Images Lignes. Application au Comptage des Passagers dans les Systèmes de Transport*, PhD thesis, INRETS, January 1997.
5 E. Oscarsson. TV-camera detecting pedestrians for traffic light control. 9th Imeko World Congress, West Berlin, May 24–28, 1982, pp. 275–282.
6 A. Mecocci, F. Bartolini, and V. Cappellini. Image sequence analysis for counting in real time people getting in and out of a bus. *Signal Processing*, 1994;35(2):105–116.
7 R. Glachet, S. Bouzar, and F. Lenoir. Estimation des flux de voyageurs dans les couloirs du métro par traitement d'images. *Recherche Transport Sécurité*, 1995; no 46:15–22.
8 L. Khoudour, L. Duvieubourg, and J.P. Deparis. Real-time passengers counting by active linear cameras. *Proceedings of SPIE*, January 29–30, 1996, San-Jose, California, USA, Vol. 2661, pp. 106–117.
9 J. Serra. *Image Analysis and Mathematical Morphology*. New York Academic Press, 1982.
10 J. Serra. Introduction to mathematical morphology. *Computer Vision, Graphics and Image Processing*, 1986;35(3):283–305.
11 D. Aubert, F. Guichard, and S. Bouchafa. Time-scale change detection applied to real time abnormal stationarity monitoring. *Real-Time Imaging*, 2004;10(1):9–22.
12 P. Maragos. A representation theory for morphological image and signal processing. *IEEE PAMI*, 1989;11(6):586–599.
13 S. Bouchafa. *Contrast-Invariant Motion Detection: Application to Abnormal Crowd Behavior Detection in Subway Corridors*, PhD dissertation, Univ. Paris VI – INRETS, 1998.
14 F. Guichard and J.M. Morel. Partial differential equations and image iterative filtering. In I.S. Puh and G.A. Watson, (Eds.), The State of the Art in Numerical Analysis, based on the Proceedings of a Conference organised by the Institute of Mathematics and its Applications (IMA), University of York, York, UK, April 1–4, 1996, (Institute of Mathematics and its Applications Conference Series New Series Vol. 63), Clarendon Press, Oxford, 1997, pp. 525–562.

Chapter 8

Tracking football players with multiple cameras

D. Thirde, M. Xu and J. Orwell

8.1 Introduction

Sports scenarios present an interesting challenge for visual surveillance applications. Here, we describe a system, and a set of techniques, for tracking players in a football (soccer) environment. The system input is video data from static cameras with overlapping fields-of-view at a football stadium. The output is the real-world, real-time positions of football players during a match. In this chapter, we discuss the problems and solutions that have arisen whilst designing systems and algorithms for this purpose.

The overall application output is the positions of players, and the ball, during a football match. This output can be used live, for entertainment – augmenting digital TV or low-bandwidth match play animations for Web or wireless display – and played back for analysis of fitness and tactics of the teams and players. At this stage, the technology is not sufficient to allow reliable recognition of the identities of each player in the game, so at present we aim only at automatic recognition of the team of each player. Indeed, there are five different uniforms of clothing on the pitch (two teams, two goalkeepers and the referees), so we aim to recognise these different categories of player.

Several research projects on tracking soccer players have published results in this field. Intille and Bobick [1] track players, using the concept of a closed-world, in the broadcast TV footage of American football games. This concept is defined as 'space–time region of an image sequence in which the complete taxonomy of objects is known, and in which each pixel should be explained as belonging to one of those objects'. The complete pitch forms the overall closed world, provided the image sequences cover sufficiently. Also, smaller, dynamically generated closed worlds are formed by the relative proximity of players' movement. For example, if two isolated players converge, then the knowledge that there are exactly two players will help to interpret their motion when they form an occluded group.

A monocular TV sequence is also the input data for References 2 and 3, in which panoramic views and player trajectories are computed. The SoccerMan [4] project analyses two synchronised video sequences of a soccer game, which are captured with pan–tilt–zoom cameras and generates an animated virtual 3D view of the given scene. These projects use one or two pan–tilt–zoom cameras to improve the image resolution of players, and the correspondence between frames has to be made on the basis of matching field lines or arcs. A monocular static camera has also been used by Needham and Boyle [5], who track the players of an indoor five-a-side soccer game with the CONDENSATION algorithm.

There exist several limitations in such single-view systems as above. First, a single camera can only cover either the whole pitch with a very low resolution for the players or part of the pitch with some players lost in the data. This is disadvantageous in players' team recognition or activity analysis of the whole team. Second, football players are frequently occluded by each other, especially when they line up towards the camera. Due to the perspective view of the camera, the dynamic occlusion may occur even if these players are far from each other on the ground plane. Third, given the ball observation in a single image, the ball position in 3D world co-ordinates is only constrained in a viewing ray and cannot be uniquely decided. This degrades the prospect of these single-view applications because the ball is always the highlight in a soccer game and may fly over the ground.

An alternative approach to improving players' resolution is to use multiple stationary cameras. Although this method requires dedicated static cameras, it increases the overall field-of-view, minimises the effects of dynamic occlusion, provides 3D estimates of ball location and improves the accuracy and robustness of estimation due to information fusion. There are different ways of using multi-view data. Cai and Aggarwal [6] and Khan *et al*. [7] track each target using the best-view camera and then hand it over to the neighbouring camera once it leaves the field-of-view of the current camera. Stein [8] and Black *et al*. [9] assume that all the targets are in the same plane (e.g. the ground plane) and compute the homography transformation between the coordinates of two overlapping images captured with uncalibrated cameras. The measurement association between cameras is made according to the homography. Another method is to calibrate the cameras and project each point in an image plane as a viewing ray in the real 3D world or to a point in a known plane (e.g. the ground plane) in the 3D world, which is able to determine the 3D real world co-ordinate with the intersecting rays from two or more cameras. This has been used in the VSAM (Video and Surveillance Monitoring) project [10], which estimates object geolocations by intersecting target viewing rays with a terrain model and hands over targets to the closest camera with pan–tilt–zoom control. Because our target application has to determine the 3D location of the ball, this method is also the one employed in the work described in this chapter.

Our system uses approximately eight or nine digital video cameras statically positioned around the stadium, and calibrated to a common ground-plane co-ordinate system using Tsai's algorithm [11]. A two-stage processing architecture is used: the details of the system architecture are provided in Section 8.2. The first processing stage is the extraction of players' information from the image streams observed by each camera, as described in Section 8.3. The data from each camera are input to a central tracking process, described in Section 8.4, to update the state estimates of the players. This includes the estimate of which of the five possible uniforms each player is wearing (two outfield teams, two goalkeepers and the three referees: in this work, 'player' includes the referees). The output from this central tracking process, at each time step, is estimated for the player positions. The tracker indicates the category (team) of each player: the identification of individual players is not attempted, given the resolution of input data, so only the team is recognised. The ball tracking methods are outside the scope of this chapter but are covered elsewhere [12].

This chapter contains a detailed description of the architecture and algorithms for estimating player positions from multiple sources. The novel components include a method for using partial observations to split grouped targets, multi-sensor data fusion to improve target visibility and tracking accuracy, and analysis of local fixed population tracking. Data fusion is performed in a common ground plane, allowing data from cameras to be added and removed from this common co-ordinate system relatively easily.

8.2 System architecture

In this section we describe the overall system, in which the algorithms described later on are implemented and integrated. As already stated, a two-tier processing architecture is proposed. The first tier is the video processing stage: the input is the image sequence from the camera, and the output is a set of 'features'. These features approximately describe the position and category of all players observed by each camera (formally, each feature consists of a 2D ground-plane position, its spatial covariance and a probabilistic estimate of the category: Section 8.3 provides details of how these features are estimated). Each camera is connected to a processing unit called a feature server, reflecting its position in the overall architecture: features are supplied to the second processing tier, called the multi-view tracker. In this second processing stage, all features from the multiple camera inputs are used to update a single model of the state of the game. This game-state is finally passed through a phase of marking-up which will be responsible for generating the XML output that is used by third party applications to deliver results to their respective target audiences.

8.2.1 *Arrangement of system components*

The topology of the system is shown in Figure 8.1: each camera is connected to a feature server (video processing module) – these are connected to the multi-view tracker, which provides the output stream. The cameras are positioned at suitable locations around the stadium and are connected to a rack of eight feature servers through a network of fibre optics as depicted in Figure 8.1. The position of the cameras is governed by the layout of the chosen stadium and the requirement to

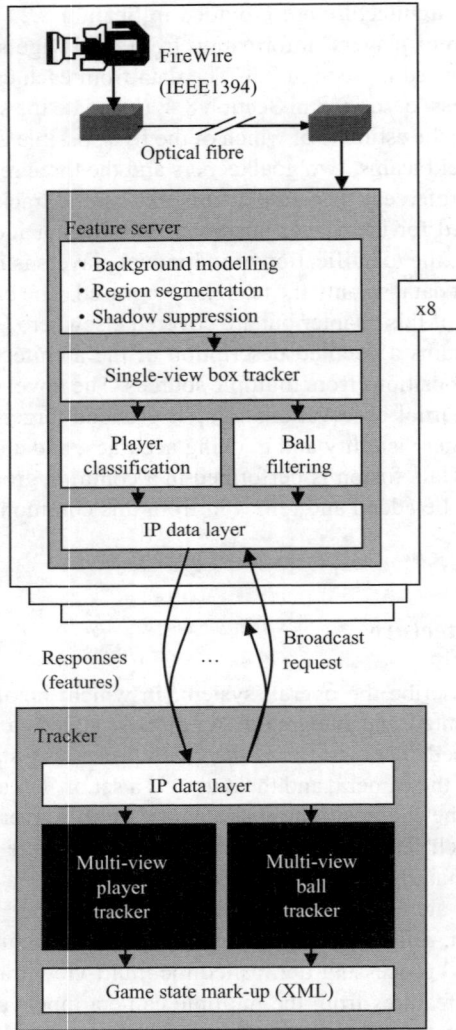

Figure 8.1   System architecture

achieve an optimal view of the football pitch: good resolution of each area, especially the goal-mouths, is more important than multiple views of each area.

For a given arrangement of camera positions, there are two strategies for placement of the processing stages. The first strategy (which is followed in our current prototype) is to house all processing components in a single location. In this case, all video data need to be transmitted to this point. Current technology dictates either analogue or digital cable is required for this: in our prototype, a combination of electrical and optical cable was used to transmit a digital video (DV) format signal. The feature servers are connected to the multi-view tracker using an IP/Ethernet network which is used to communicate the sets of features viewed from each camera.

This configuration of physical location of components is influenced by the requirement of minimising the profile of the installations above the stadium. If that requirement were not so important, the overall bandwidth requirements could be considerably reduced by locating the feature servers alongside the cameras. Then, only the feature sets would need transporting along the long distances from each camera to the site of the multi-view tracker: this could be achieved with regular or even wireless Ethernet, rather than the optic fibre needed for the video data.

8.2.2   Synchronous feature sets

The input requirements for the algorithms naturally affect the design decisions of the system architecture. For the multi-view tracker, an important consideration is whether its input feature sets need to be synchronous or whether asynchronous inputs are sufficient. For the latter case, the data association and fusion steps of the multi-view tracker are significantly more complicated; so the stronger requirement of 'synchronous' feature sets was imposed. This term is used somewhat loosely, as it was not practical to exactly synchronise the phase of the 25 Hz DV cameras, so the system is granted a tolerance of at most one frame of input (i.e. 40 ms). A secondary requirement is that, for any given time step, the interval between successive feature sets will be the same multiple of 40 ms (e.g. 80 or 120 ms) for all cameras. That interval will be determined by whichever feature server needs the longest processing time: this may depend on the content, such as sudden lighting changes or lots of players in its field of view.

To satisfy the synchronisation requirements outlined above, any proposed method must accommodate the variability of the processing time taken by each feature server; the method should also be robust to catastrophic failure in which one or more feature servers fail to provide any data at all. Two possible methods are, first pre-arranged time intervals, scheduled by pre-synchronised clocks; and second a network-based request for the data, to the feature servers, scheduled whenever all feature servers are available to respond to this request, all with the same time interval since they last responded.

We chose the 'request–response' method, since there appeared no easy way to calculate the optimal time-step for any such pre-arranged interval. The multi-view tracker publishes the request when it is ready to receive input. To satisfy the secondary requirement (that all inputs from all cameras have the same time-interval), the request is delayed until exactly the next whole multiple of 40 ms since the last request was published. (Even then, the requirement is not guaranteed to be satisfied, but the rate of error sustained though variabilities of the UDP and TCP systems is estimated to be insignificant.)

Thus, the multi-view tracker (second stage) is responsible for synchronisation of video-processing feature servers (first stage). Each iteration (or frame) of the process comprises a single (broadcast) request issued by the tracker at a given time. The feature servers then respond by taking the latest frame in the video stream, processing it and transmitting the resultant features back to the tracker. Synchronisation of the feature sets is implied as the multi-view tracker will record the time at which the request was made.

The request–response action repeats continually: another request is issued as soon as all feature sets have been received. In parallel, these feature sets are passed on to be processed by components running inside the multi-view tracker. The results of this process are then marked up (into XML) and delivered through a peer-to-peer connection with any compliant third party application.

One possible benefit of the chosen method is that the request could be used as a vehicle to send information to the first processing stage from the second processing stage. For example, observations from one camera may help interpret a crowded scene in another camera: these could be sent via the request signal. However, this facility is not used in the algorithms described in the sections below.

8.2.3   Format of feature data

The second stage, multi-view tracker process does not have access to video data processed in the first stage (the video processing). Therefore, the feature data must include all that is required by the second stage components to generate reliable estimates of the position for the people (and ball) present in the scene. The composition of the feature is thus dictated by the requirements of the second stage process. The process described in Section 8.4 requires the bounding box, estimated ground-plane location and covariance, and the category estimate (defined as a seven-element vector, the elements summing to 1, and corresponding to the five different uniforms, the ball and 'other'). Further information is included, for example, the single-view tracker ID tag, so that the multi-view tracker can implement target-to-target data association [13]. Common software development design patterns are used to manage the process of transmitting these features to the tracker hardware by managing the process of serialising to and from a byte stream (UDP socket). This includes the task of ensuring compatibility between different platforms.

8.3   Video processing

The feature server uses a three-step approach: foreground detection, single-view tracking and category classification. This is indicated in Figure 8.1, and described in the following three sub-sections. As discussed in Section 8.2.3, the required output from this stage is a set of features, each consisting of a 2D ground-plane position, its spatial covariance and a probabilistic category estimate.

8.3.1   Foreground detection

The first step is moving object detection based on image differencing, the output of which is connected foreground regions and region-based representations such as centroids, bounding boxes and areas. (This output must be segmented into the individual players (Section 8.3.2) and classified into the possible categories (Section 8.3.3) before it is a completely specified feature.) The basic method for using Gaussian mixtures to model the background is presented in Section 8.3.1.1; the update rule is given in Section 8.3.1.1.3. To save processing time and reduce the number of false alarms, two masks are constructed and used to gate the output of the image-differencing operation. The pitch mask, M_p, models the pitch to exclude movements in the crowd: its construction is described in Section 8.3.1.1.1. The field-line mask, M_f, helps remove false alarms caused by a shaking camera; this is described in Section 8.3.1.1.2.

8.3.1.1 Initialising the background images

Each pixel of an initial background image is modelled by a mixture of Gaussians [14], $(\mu_k^{(l)}, \sigma_k^{(l)}, \omega_k^{(l)})$, and learned beforehand without the requirement of an empty scene, where $\mu_k^{(l)}, \sigma_k^{(l)}$ and $\omega_k^{(l)}$ are the mean, root of the trace of the covariance matrix and weight of the lth distribution at frame k. The distribution matched to a new pixel observation $\mathbf{I}_k$ is updated as

$$\mu_k = (1 - \rho)\mu_{k-1} + \rho\mathbf{I}_k,$$
$$\sigma_k^2 = (1 - \rho)\sigma_{k-1}^2 + \rho(\mathbf{I}_k - \mu_k)^T(\mathbf{I}_k - \mu_k).$$

(8.1)

with an increased weight, where ρ is the updating rate and $\rho \in (0, 1)$. For unmatched distributions, the parameters remain the same but the weights decrease. The initial background image is selected as the distribution with the greatest weight at each pixel.

8.3.1.1.1 The pitch mask

The initial background image is first used to extract a pitch mask for avoiding false alarms from spectators and saving processing time. The pitch mask is defined as the intersection of colour-based and geometry-based masks, as shown in Figure 8.2. For the colour-based mask, the pitch region is assumed to have an almost uniform colour (though not necessarily a green colour) and occupy a dominant area of the background image. To cope with shadows on the grass, the initial background image is transformed from the RGB space to HSI space. For those pixels with moderate intensities, a hue histogram is generated, and its highest peak corresponds to the grass. After a low-pass filtering on the hue histogram, a low threshold H_L and a high threshold H_H are searched from the highest peak and decided once the histogram value decreases to 10% of the highest. Then those image pixels (u, v) within the thresholds undergo a morphological closing operation (denoted by •), with square structuring element B_1, to fill the holes caused by outliers. This results in a colour-based pitch mask:

$$M_c = \{(u, v) | H(u, v) \in [H_L, H_H]\} \bullet B_1.$$

(8.2)

On the other hand, there might exist grass-like colours outside the pitch, which can be inhibited by a geometry-based pitch mask M_g with the well-known pitch geometry back-projected onto the image plane. Suppose that E is the co-ordinate transformation from the image plane to the ground plane and P represents the co-ordinate range of the pitch on the ground plane, the overall pitch mask M_P is the intersection of both the colour-based mask M_c and geometry-based mask M_g, as illustrated

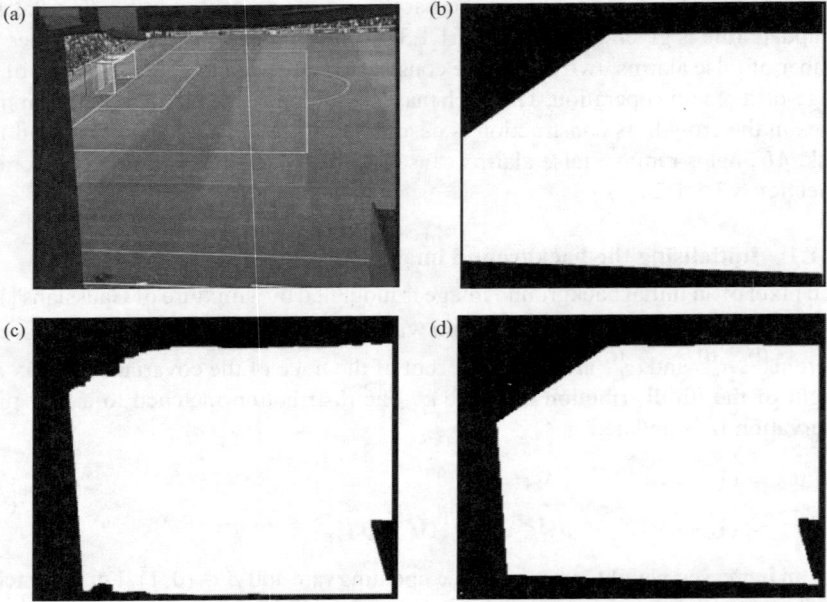

Figure 8.2   Extraction of pitch masks: (a) background image; (b) geometry-based mask; (c) colour-based mask; and (d) overall mask

in Figure 8.2:

$$M_g = \{(u, v)|E(u, v) \in P\},$$
$$M_P = M_g \cap M_c.$$

$$(8.3)$$

8.3.1.1.2   The field line mask

Another application of the initial background is to extract a field line mask for reducing the effect of camera shakes, from which many visual surveillance systems have suffered. It is observed that the edges with abrupt intensity variation are more vulnerable to camera shake. A map of such edge areas can be used to mask the false alarms caused by small-scale camera shakes. Because most of the sharp edges occur at field lines in the domain of the football pitch, this edge map is called the field line mask.

An initial field line mask is extracted on the basis of the edge response:

$$M_L = \begin{cases} 1 & \text{if } \|I(u, v) * \nabla G(u, v, s)\| < T_e, \\ 0 & \text{otherwise.} \end{cases}$$

$$(8.4)$$

The edge response is the convolution of the image intensity I with the first derivative of the Gaussian of standard deviation s. This field line mask then undergoes a morphological dilation with square structuring element B_2 to cope with the magnitude of camera shakes:

$$M_L = M_L \oplus B_2. \tag{8.5}$$

8.3.1.1.3 Background updating

Given the input image $\mathbf{I}_k$, the foreground binary map F_k is decided by comparing $\|\mathbf{I}_k - \boldsymbol{\mu}_{k-1}\|$ against a threshold. The background image is then updated using the quicker running average algorithm:

$$\boldsymbol{\mu}_k = [\alpha_L \mathbf{I}_k + (1 - \alpha_L)\boldsymbol{\mu}_{k-1}]F_k + [\alpha_H \mathbf{I}_k + (1 - \alpha_H)\boldsymbol{\mu}_{k-1}]\bar{F}_k, \tag{8.6}$$

where $0 < \alpha_L \ll \alpha_H \ll 1$. This method slowly updates the background image even in the foreground regions so that any mistake in the initial background estimate and later variations of the background will not be locked. The run-time simplification of the background model aims to accelerate the computation at the cost of some adaptivity.

The foreground binary map F_k is filtered by the pitch and field line masks to remove false alarms:

$$F_k = F_k \cap M_P \cap M_L. \tag{8.7}$$

After a connected component analysis which transforms a foreground pixel map into a foreground region map, F_k is subject to another morphological closing operation with square structuring element B_3 to fill the holes in foreground regions and bridge the gaps of splitting foreground targets caused by the field line mask. This method efficiently reduces the effect of camera shake while keeping most players intact. The foreground regions smaller than a size threshold are considered as false alarms and ignored. An example of the foreground detection is shown in Figure 8.3(a).

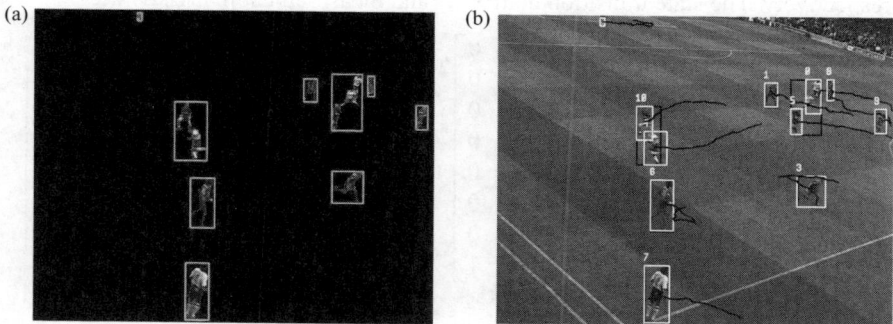

Figure 8.3   *Foreground detection (a) and single-view tracking (b). Black and white rectangles in the right image represent foreground regions and estimates, respectively*

8.3.2 Single-view tracking

The second step is a local tracking process to split features of grouped players. In the context of single camera, a group is defined as those players sharing a single connected component of the foreground image. Players are frequently grouped, even when they are far from each other on the ground plane, due to the perspective view of the camera. For such a group of players, estimation of their features (position and category) is not as straightforward as it is for the isolated player. The isolated player is fully observed since all bounding box edges refer to his position. The method described below derives the measurements for grouped players, using the partial observations available from a bounding box that contains more than one player.

8.3.2.1 Tracking model

The image-plane position of each player is modelled with their bounding box and centroid co-ordinates and represented by a state $\mathbf{x}_I$ and measurement $\mathbf{z}_I$ in a Kalman filter:

$$\mathbf{x}_I = [r_c \quad c_c \quad \dot{r}_c \quad \dot{c}_c \quad \Delta r_1 \quad \Delta c_1 \quad \Delta r_2 \quad \Delta c_2]^T,$$

$$\mathbf{z}_I = [r_c \quad c_c \quad r_1 \quad c_1 \quad r_2 \quad c_2]^T,$$

(8.8)

where (r_c, c_c) is the centroid, $(\dot{r}_c, \dot{c}_c)$ is the velocity and r_1, c_1, r_2, c_2 represent the top, left, bottom and right bounding edges, respectively ($r_1 < r_2$ and $c_1 < c_2$). $(\Delta r_1, \Delta c_1)$ and $(\Delta r_2, \Delta c_2)$ are the relative positions of the two opposite bounding box corners to the centroid.

The image-plane process evolution and measurement equations are

$$\mathbf{x}_I(k+1) = \mathbf{A}_I \mathbf{x}_I(k) + \mathbf{w}_I(k),$$

$$\mathbf{z}_I(k) = \mathbf{H}_I \mathbf{x}_I(k) + \mathbf{v}_I(k),$$

(8.9)

where $\mathbf{w}_I$ and $\mathbf{v}_I$ are the image-plane process noise and measurement noise, respectively. The state transition matrix $\mathbf{A}_I$ and measurement matrix $\mathbf{H}_I$ are

$$\mathbf{A}_I = \begin{bmatrix} 1 & 0 & T & 0 & 0 & 0 & 0 & 0 \\ 0 & 1 & 0 & T & 0 & 0 & 0 & 0 \\ 0 & 0 & 1 & 0 & 0 & 0 & 0 & 0 \\ 0 & 0 & 0 & 1 & 0 & 0 & 0 & 0 \\ 0 & 0 & 0 & 0 & 1 & 0 & 0 & 0 \\ 0 & 0 & 0 & 0 & 0 & 1 & 0 & 0 \\ 0 & 0 & 0 & 0 & 0 & 0 & 1 & 0 \\ 0 & 0 & 0 & 0 & 0 & 0 & 0 & 1 \end{bmatrix} \quad \mathbf{H}_I = \begin{bmatrix} 1 & 0 & 0 & 0 & 0 & 0 & 0 & 0 \\ 0 & 1 & 0 & 0 & 0 & 0 & 0 & 0 \\ 1 & 0 & 0 & 0 & 1 & 0 & 0 & 0 \\ 0 & 1 & 0 & 0 & 0 & 1 & 0 & 0 \\ 1 & 0 & 0 & 0 & 0 & 0 & 1 & 0 \\ 0 & 1 & 0 & 0 & 0 & 0 & 0 & 1 \end{bmatrix}.$$

(8.10)

As in Reference 15, it is assumed that each target has a slowly varying height and width. Once some bounding edge of a target is decided to be observable, its opposite, unobservable bounding edge can be roughly estimated. The decisions about which targets are grouped, and which bounding edges are observable, are based on the relative positions of the foreground region and the predictions of individual targets. If the predicted centroids of multiple targets are within the bounding box of the same foreground region, then these targets are considered to be in group. Each predicted bounding edge of a target, that is outermost in its group, is treated as 'observed', and associated with the corresponding edge of the bounding box that represents the group of players. The centroids of each player, and the remaining internal edges of the individuals' bounding boxes, are not observed at this time step. For these variables, the corresponding elements in the covariance matrices are increased so that they contribute little to the estimation process. Therefore, the estimate depends more on the observable variables. Because the estimate is updated using partial measurements whenever available, it is more robust and accurate than using prediction only. An example of single-view tracking with two groups of players is shown in Figure 8.3(b).

8.3.2.2 Track correction

Football players are agile targets, with abrupt changes in movement and hence do not fit the constant velocity or constant acceleration dynamic model particularly well. The estimation errors have to be corrected periodically to avoid losing the player through an accumulating error. For each tracked player, once the estimated centroid is found outside any foreground region, a tracking correction process will be triggered, in which the closest foreground region to the estimated centroid will be searched in a horizontal scan (Figure 8.4). If this foreground region along the scan line is not wider than the estimated width of the player, the corrected centroid will be located in the middle of the foreground region (Figure 8.4(a)). Otherwise, it will be relocated according to the closest edge of the foreground region and the estimated width of the player (Figure 8.4(b)).

8.3.2.3 Expressing the features in ground plane co-ordinates

For an isolated player, the image measurement comes from the foot of foreground region directly, which prevents estimation errors accumulating in a hierarchy of Kalman filters. As foreground detection in an image is a pixel-wise operation, we

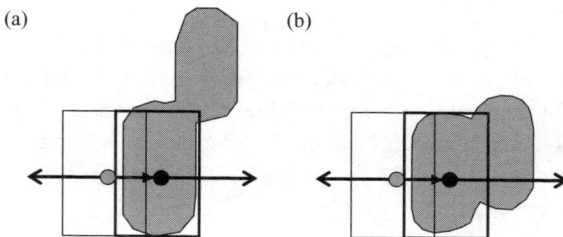

(a)   (b)

Figure 8.4   *Tracking correction. Grey blobs represent foreground regions, rectangles in thin lines represent initial estimate and those in thick lines represent corrected estimate*

estimate the measurement covariance in an image plane is a constant and diagonal matrix Λ. The corresponding measurement and covariance projected on the ground plane from the ith image plane are

$$\mathbf{z}^{(i)} = E^{(i)}(r_2, c_c),$$
$$\mathbf{R}^{(i)} = \mathbf{J}_E^{(i)}(r_2, c_c)\Lambda\mathbf{J}_E^{(i)}(r_2, c_c)^{\mathrm{T}},$$

(8.11)

where $E^{(i)}$ is the coordinate transformation from the ith image plane to the ground plane and $\mathbf{J}_E^{(i)}$ is the Jacobian matrix of $E^{(i)}$. These measurements and their covariances are demonstrated in Figure 8.9(b) and (c).

For players that are part of a group, the inevitable increase in uncertainty about their position is modelled with a scaling factor $\lambda > 1$ for the covariance matrix Λ:

$$\mathbf{z}^{(i)} = E^{(i)}(\hat{r}_2, \hat{c}_c),$$
$$\mathbf{R}^{(i)} = \mathbf{J}_E^{(i)}(\hat{r}_2, \hat{c}_c)\lambda\Lambda\,\mathbf{J}_E^{(i)}(\hat{r}_2, \hat{c}_c)^{\mathrm{T}}.$$

(8.12)

The bounding box tracking with partial observations is the most effective for groups of two targets. With more than two players in a group, the number of unobserved components for each target renders the technique less effective, and the target uncertainty increases. Therefore, for the groups of more than two targets, the foreground region is treated as a single measurement and forwarded to the multi-view tracking stage to resolve into the separate players.

8.3.3   Category classification

The final step of the video processing stage adds to each measurement a probabilistic estimation of the category (i.e. the player's uniform). Histogram intersection [16] is a technique for classifying detected players that overcomes the problem of distracting background pixels; it has been applied to locating and tracking non-rigid objects within a scene in References 17 and 18. For each player we wish to extract a five-element vector $\mathbf{c}_j^{(i)}(k)$, indicating the probability that the feature is a player wearing one of the five categories of uniform. This probability is calculated as the product of a prior probability of observing a player at a given position on the pitch, and a conditional likelihood that an observation represents a player of that category.

To compute this conditional likelihood, we use a classifier based on RGB histogram intersections. For each detected player, we extract and normalise its histogram, h, that ranges jointly over the three RGB channels. The histogram intersection F is calculated between this and the five model histograms g_c ($c \in [1, 5]$). This is defined to be

$$F(h, g_c) = \sum_b \min(h(b), g_c(b)),$$

(8.13)

where b ranges over all elements in the histogram. This function is evaluated between each player and the five model histograms to estimate a measure of the likelihood that the observation represents a player of that category.

The *posteriori* probability is a product of the above likelihood with a prior probability. For outfield players, this is uniform, that is, they are equally probable over the entire pitch. However, the goalkeepers and linesmen are heavily constrained. The prior is only non-zero in and around the penalty area for the goalkeepers and along the touch lines for linesmen.

The method of generating the model histograms is described below. We adopt a semi-supervised strategy where an operator labels examples of player observations before the match starts. From this dataset we can build an initial estimate g'_c of the model histogram for each category specified. However, this estimate also includes background noise alongside the player observations. To reduce the effect of this background noise we calculate the common histogram intersection, S, of the five model histograms:

$$S(b) = \min(g'_1(b), \ldots, g'_5(b)). \tag{8.14}$$

This 'noise floor' is then removed from the initial estimate, to provide a refined estimate for each model c:

$$g_c(b) = g'_c(b) - S(b). \tag{8.15}$$

This refined estimate is used in Equation (8.13) to compute $F(h, g_c)$ – the similarity between the observation and each of the category models. They are scaled to a unit sum, and interpreted as a distribution estimating the probability that the observation feature represents a player of category c. These data, alongside the position estimates, are forwarded to the second stage of processing, described in Sections 8.4 and 8.5.

8.4   Multi-view, multi-person tracking

This section describes the first two steps of the second stage of processing. This stage is that in which sets of measurements from the cameras are integrated into tracks: a unified representation of the player positions over time. It has a structure similar to that of the first stage: a succession of steps that add increasing levels of detail and cohesion to the representation. The first two steps, described in this section, implement an unconstrained tracking process. The first step is to associate measurements of the feature servers to established targets and update the estimates of these targets. The second step is to initialise targets for the measurements unmatched to any existing target. The results are then forwarded to a selection process, described in Section 8.5, in which the constraints applicable to each category of players (ten outfield players and one goalkeeper per team, three referees) are used to label the output of the method described below.

8.4.1   *Associating targets and features*

Each player is modelled as a target x_t, represented by the state $\mathbf{x}_t(k)$, the state covariance $\mathbf{P}_t(k)$ and a category estimate $\mathbf{e}_t(k)$ at frame k. The state estimate is updated, if possible, by a measurement m_t fused from at least one camera. The m_t comprises a ground-plane position $\mathbf{z}_t(k)$, a covariance $\mathbf{R}_t(k)$ and a category measurement $\mathbf{c}_t(k)$. The state and measurement variables in a ground-plane Kalman filter are

$$
\begin{aligned}
\mathbf{x} &= [x \quad y \quad \dot{x} \quad \dot{y}]^{\mathrm{T}}, \\
\mathbf{z} &= [x \quad y]^{\mathrm{T}},
\end{aligned}
\tag{8.16}
$$

where (x, y) and $(\dot{x}, \dot{y})$ are the ground-plane coordinate and velocity.

The ground-plane process evolution and measurement equations are

$$
\begin{aligned}
\mathbf{x}(k+1) &= \mathbf{A}_w \mathbf{x}(k) + \mathbf{w}_w(k), \\
\mathbf{z}(k) &= \mathbf{H}_w \mathbf{x}(k) + \mathbf{v}_w(k),
\end{aligned}
\tag{8.17}
$$

where $\mathbf{w}_w$ and $\mathbf{v}_w$ are the ground-plane process noise and measurement noise, respectively. The state transition and measurement matrices are

$$
\mathbf{A}_w = \begin{bmatrix} 1 & 0 & T & 0 \\ 0 & 1 & 0 & T \\ 0 & 0 & 1 & 0 \\ 0 & 0 & 0 & 1 \end{bmatrix} \quad \mathbf{H}_w = \begin{bmatrix} 1 & 0 & 0 & 0 \\ 0 & 1 & 0 & 0 \end{bmatrix}.
\tag{8.18}
$$

The established targets $\{x_t\}$ are associated with the measurements $\{m_j^{(i)}\}$ from the ith camera, the result of which is expressed as an association matrix $\beta^{(i)}$. Each element $\beta_{jt}^{(i)}$ is 1 for association between the tth target and jth measurement or 0 otherwise. The association matrix is decided according to the Mahalanobis distance d_{jt} between the measurement and the target prediction:

$$
\begin{aligned}
\mathbf{S}_{jt}(k) &= \mathbf{H}_w \mathbf{P}_t(k|k-1) \mathbf{H}_w^{\mathrm{T}} + \mathbf{R}_j^{(i)}(k), \\
d_{jt}^2 &= [\mathbf{z}_j^{(i)}(k) - \mathbf{H}_w \hat{\mathbf{x}}_t(k|k-1)]^{\mathrm{T}} \mathbf{S}_{jt}(k)^{-1} [\mathbf{z}_j^{(i)}(k) - \mathbf{H}_w \hat{\mathbf{x}}_t(k|k-1)].
\end{aligned}
\tag{8.19}
$$

For a possible association this distance must be within a validation gate. Then the nearest neighbour algorithm is applied and each target can be associated with at most one measurement from a camera, that is, $\sum_j \beta_{jt}^{(i)} \leq 1$. The underlying assumption is that a player can give rise to at most one measurement in a camera. $\sum_j \beta_{jt}^{(i)} = 0$ corresponds to missing measurement cases. On the other hand, no constraints are imposed on $\sum_t \beta_{jt}^{(i)}$, the number of targets matched to a measurement. Each measurement may be assigned to one target, multiple targets (for merged players) or no target (for false alarms).

The individual camera measurements assigned to each target are weighted by measurement uncertainties and integrated into an overall measurement as follows:

$$\mathbf{R}_t = \left[\sum_i \sum_j \beta_{jt}^{(i)} \left(\mathbf{R}_j^{(i)}\right)^{-1} \right]^{-1}, \tag{8.20}$$

$$\mathbf{z}_t = \mathbf{R}_t \left[\sum_i \sum_j \beta_{jt}^{(i)} \left(\mathbf{R}_j^{(i)}\right)^{-1} \mathbf{z}_j^{(i)} \right], \tag{8.21}$$

$$\mathbf{c}_t = \sum_i w_t^{(i)} \sum_j \beta_{jt}^{(i)} \mathbf{c}_j^{(i)}, \tag{8.22}$$

$$w_t^{(i)} = \frac{\sum_j \beta_{jt}^{(i)} / \mathrm{tr}\left(\mathbf{R}_j^{(i)}\right)}{\sum_i \sum_j \beta_{jt}^{(i)} / \mathrm{tr}\left(\mathbf{R}_j^{(i)}\right)}, \tag{8.23}$$

where tr() represents the trace of a matrix. Although written like a summation, items like $\sum_j \beta_{jt}^{(i)} f(m_j^{(i)})$ in Equations (8.20)–(8.23) is actually a single measurement, from the ith camera, assigned to the tth target due to $\sum_j \beta_{jt}^{(i)} \leq 1$. Equation (8.20) indicates that the fused measurement is more accurate than any individual measurement, while Equation (8.21) indicates that the fused measurement is biased to the most accurate measurement from individual cameras.

Each target with a measurement is then updated using the integrated measurement:

$$\mathbf{K}_t(k) = \mathbf{P}_t(k|k-1)\mathbf{H}_w^{\mathsf{T}}[\mathbf{H}_w\mathbf{P}_t(k|k-1)\mathbf{H}_w^{\mathsf{T}} + \mathbf{R}_t(k)]^{-1} \tag{8.24}$$

$$\hat{\mathbf{x}}_t(k|k) = \hat{\mathbf{x}}_t(k|k-1) + \mathbf{K}_t(k)[\mathbf{z}_t(k) - \mathbf{H}_w\hat{\mathbf{x}}_t(k|k-1)] \tag{8.25}$$

$$\mathbf{P}_t(k|k) = [\mathbf{I} - \mathbf{K}_t(k)\mathbf{H}_w]\mathbf{P}_t(k|k-1) \tag{8.26}$$

$$\hat{\mathbf{e}}_t(k) = (1-\eta)\hat{\mathbf{e}}_t(k-1) + \eta\mathbf{c}_t(k), \tag{8.27}$$

where $0 < \eta < 1$.

If no measurement is available for an existing target, then the state estimate is updated using its prior estimate only. In this case the state covariance increases linearly with time. Once a target has no measurement over a certain number of frames, the state covariance reflecting uncertainty will be larger than a tolerance. The tracking of this target will be automatically terminated and this target will be re-tracked as a new one when the measurement flow is resumed.

8.4.2 Target initialisation

After checking measurements against existing targets, there may be some measurements unmatched. Then those measurements, each from a different camera, are checked against each other to find potential new targets. If there exists an unmatched measurement $\mathbf{z}_{j_1}^{(i_1)}$ from the i_1th camera, then a new target x_n will be established. All the association matrices $\beta^{(i)}$ are extended by one column, each element $\beta_{jn}^{(i)}$ of which indicates the correspondence between the new target and the jth measurement from the ith camera. For the i_1th camera, the measurement $z_{j_1}^{(i_1)}$ is automatically associated with the new target:

$$\beta_{jn}^{(i_1)} = \begin{cases} 1 & \text{if } j = j_1, \\ 0 & \text{otherwise.} \end{cases} \tag{8.28}$$

For each unmatched measurement from the other cameras, $z_j^{(i)}$ with covariance $R_j^{(i)}$, it is checked against the measurement $z_{j_1}^{(i_1)}$ with $R_{j_1}^{(i_1)}$, and thought to be associated with the new target if the Mahalanobis distance,

$$d_{i_1 i}^2 = \left[z_j^{(i)}(k) - z_{j_1}^{(i_1)}(k)\right]^T \left[R_j^{(i)}(k) + R_{j_1}^{(i_1)}(k)\right]^{-1} \left[z_j^{(i)}(k) - z_{j_1}^{(i_1)}(k)\right], \tag{8.29}$$

is within a validation gate and smallest in all the unmatched measurements from the ith camera. Therefore, the new target can only be associated with at most one measurement from each camera. All the individual camera measurements assigned to the new target are then integrated into an overall measurement, z_n, R_n and c_n, using Equations (8.20)–(8.23) and substituting the subscript t with n. The new target is then initialised with the integrated measurement:

$$\hat{x}_n(k|k) = \left[z_n(k)^T \quad 0 \quad 0\right]^T, \tag{8.30}$$

$$P_n(k|k) = \begin{bmatrix} R_n(k) & 0_2 \\ 0_2 & \sigma_v^2 I_2 \end{bmatrix}, \tag{8.31}$$

$$\hat{e}_n(k) = c_n(k), \tag{8.32}$$

where 0_2 and I_2 are 2×2 zero and identity matrices, respectively.

In the initialisation scheme as above, any false alarm in a single camera will result in a new target being established. A discrete alternative, as implemented in our system, is to check the number of supported cameras against the expected number of supported cameras for each set of measurements that are not matched to any existing targets but associated to each other. The expected number of supported cameras can be obtained by counting the fields-of-view within which the measurements are located. A temporal target will be established only if these two numbers are consistent. Therefore, only when players are well separated and observed can they be recognised as temporal targets. The temporal targets will become new targets once they have been tracked for a number of frames.

8.5   Strategies for selection and classification of tracks

The multi-view tracker, described in Section 8.4, maintains an unconstrained set of model targets. The estimate of each target state is represented by a pdf for its position (a mean and covariance); and a probability distribution for its category (a seven-element vector comprising the five team categories, the ball and the null category). In many tracking applications, the overall tracker output would be directly implied by these, that is, for each tracked target, output its most likely position and category, as solely determined by its own model. However, this unconstrained output would often result in the wrong number of players in each team. That error could be caused by several reasons, manifesting in either the wrong number of targets in the model and/or the wrong estimate of category in one or more of the target models. At this stage it is useful to re-examine the application requirements to consider how best to handle possible errors and uncertainty in the model of player positions. This is undertaken in Section 8.5.1, with the conclusion that it is an application require-ment to output the correct number of players, even though, by some error metrics, it may decrease the apparent accuracy of the system. In Section 8.5.2 a framework is presented for satisfying the overall population constraint, aiming for the maxi-mum *a posteriori* (MAP) likelihood of labelling of targets, given their individual probability distributions. In Section 8.5.3 we provide some simple heuristics for the estimation of these probabilities, and a sub-optimal (non-exhaustive) technique for using these to satisfy a given fixed population constraint. The results achieved with this technique leave plenty of scope for improvement, as the probability estimates are not very robust to occlusions, resulting in frequent discontinuities in the assigned labels. In Section 8.5.4 we introduce a strategy to combat this uncertainty. Here, we aim to exploit the temporal continuity of the players, to partition them into groups, each group satisfying a maximum spanning distance criterion. Each group has its own local population constraint: the MAP likelihood procedure can now be applied to each group separately. This results in an improvement of the track continuity and accuracy. The introduction of this method raises many further questions, which are discussed in Section 8.5.5.

8.5.1 Application requirements

In this section we consider what constraints to impose on the output of the multi-view tracker. In this context, an unconstrained output would simply be that produced by the process described in Section 8.4: observations are associated with targets, which are updated, created and deleted accordingly. Each has a category estimate, which is a weighted sum of the estimates from the individual measurements. In many other applications where the total number of people is not known, this may be the most appropriate form of output.

However, for our application, the total number of people on the pitch is a known quantity. The default number is 25: ten outfield players and one goal-keeper per team, and the referee with two assistants. If a red card is issued then this num-ber is decremented; other departures from the default include players temporarily leaving the field for injury; medical staff or supporters entering the field; and sub-stitutes warming-up on the edge of the field. The dismissal of a player is a key event, and the system must have reliable access to the number of players currently engaged in the game: we assume that news of this event will be available from some reliable external source. The other cases are less important: they could be handled with either a manual real-time input of the current number of people on the play-field or a system robust enough to automatically classify these targets into the 'null' category.

We could require that our system constrain the output so that it comprises the correct number of players from each category. An analogous constraint in the domain of optical character recognition is postcode recognition: a system could be constrained to output only valid postcodes. An output containing the wrong number of players is not permitted, just as an invalid sequence of letters in a postcode may be rejected. The model can be constrained globally, as shown in Figure 8.5, on the constraints can be locally imposed, as shown in Figure 8.7 and described in Section 5.4.

8.5.2   The MAP labelling hypothesis

The labelling of the targets in the model involves selection (recognising which tracks are the genuine ones, when there are more targets in the model than players on the field) and classification (recognising which target belongs to which category).

In this section we describe a methodology for allocating the fixed population of players and referees using an MAP labelling approach. This is shown schematically in Figure 8.5. We describe a method for labelling each target in the model with one of six categories, subject to the constraint that a label of category c can only be used N_c times. In all cases we also use the null category ($c = 6$); the population of this category is unconstrained, that is, the null population is equal to the number of tracks minus the number of fixed population targets (provided that there is a surplus of tracks).

To select the tracks, we define a corresponding label vector $\mathbf{q}_t$, composed of six binary elements. Setting $q_t^c(k) = 1$ represents a decision to assign one label from

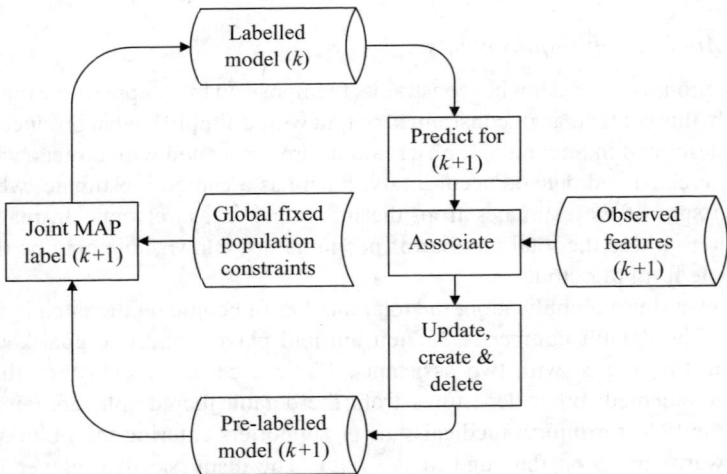

Figure 8.5   *Flow chart for the tracking process with a global fixed population constraint. The 'labelled' model at frame k is updated with observations, to obtain a pre-labelled model for frame (k + 1). The global fixed population constraint is re-imposed in a joint maximum a priori (MAP) labelling process*

category c to track t at frame k. There are two label constraints: there must be at most one label per track, and the total label assignments must satisfy the fixed population constraint. Given the fixed population total in each category N_c we have

$$\forall c \sum_t q_t^c = N_c. \tag{8.33}$$

There is no constraint on N_6 – the number of false alarms in the null category. Likewise, a vector $\mathbf{p}_t(k)$ is used to represent the probability distribution for each track, composed of the six elements, $p_t^c, c = \{1, \ldots, 6\}$ representing the probability that track t represents category C. It is assumed that these outcomes are exhaustive (i.e. the track principally represents exactly one or zero players) and so these elements always sum to 1.

Assuming the probability estimates are independent, the likelihood of all labels being correct is the product of those elements picked out by the label vectors:

$$p(\{\mathbf{q}\}) = \Pi_t \mathbf{p}_t \mathbf{q}_t. \tag{8.34}$$

This method therefore requires a means of estimating the $\{\mathbf{p}_t\}$, and a procedure for setting the elements of the $\{\mathbf{q}_t\}$ to obey the fixed population constraint and maximise the total *a posteriori* probability.

8.5.3   Estimates for target categories

First we describe the method for calculating the $\{\mathbf{p}_t(k)\}$. A likelihood $\{\mathbf{p}_t'(k)\}$ has three component factors, which are then combined with a prior composed from $\{\mathbf{p}_t(k-1)\}$ and $\{\mathbf{q}_t(k-1)\}$.

The first factor is the probability the track represents any player, rather than a false alarm of any kind. The second factor is the probability that it represents a player of category c (given that it represents a player). Finally, we incorporate the probability that the track is the principal representation of this player, that is, it is not the case that another track represents this particular player better. The first factor incorporates both the track age (in frames) k_1, and the number of frames k_2 since the last observation. These variables are used in an exponential model of the probability of a false alarm, using κ_1 and κ_2 to scale k_1 and k_2 appropriately. The second factor is simply the category estimate from the associated observations, $\mathbf{c}_t(k)$. These are combined as

$$p_t'(k) = \exp^{-k_2/\kappa_2}(1 - \exp^{-k_1/\kappa_1})\mathbf{c}_t(k). \tag{8.35}$$

The third factor is intended to remedy situations where a single player is represented by two tracks. The problem is addressed here by counting up $N^c(k)$, the number of tracks most likely to be of category c. If this is greater than N_c (the fixed population of that category), then the $N^c(k) - N_c$ such tracks closest to another target of the same category are estimated to be duplicate tracks, and considered to be false alarms. This is effectively a filter of tracks superfluous to a given category. If there are fewer targets than the fixed population of a given category, then the constraint that there be at most one label per track is relaxed (however, we have found in practice this to be a very rare occurrence).

Using a per frame track selection policy causes temporal discontinuity of selected tracks leading to the situation where previously confident and mature tracks can be replaced by newer or anomalous tracks. In particular, probability estimates given the fixed population constraint can be incorporated via inclusion of the previous label result. To incorporate this prior knowledge we use a recursive filter to update the overall probability estimates, using a joint rolling average of the form

$$\mathbf{p}_t(k) = \gamma \omega \mathbf{p}_t(k-1) + \gamma(1-\omega)\mathbf{q}_t(k-1) + (1-\gamma)\mathbf{p}'_t(k), \qquad (8.36)$$

where γ and ω are weights that control the update rate for the matrix based on the previous frame matrix $\mathbf{p}_t(k-1)$, the previous frame label matrix $\mathbf{q}_t(k-1)$ and the current frame observed matrix $\mathbf{p}'_t(k)$. After the $\mathbf{p}_t(k)$ are evaluated, the $\mathbf{q}_t(k)$ are assigned by choosing the most probable elements first, until the population constraint is satisfied.

8.5.4   Local fixed population constraints

The above method for allocating labels to tracks is sensitive to errors in both the identification of the players (i.e. to which team they belong) and also the total number of tracks. While it would certainly be useful to reduce the cause of these errors, for example, by introducing more robust colour analysis techniques, this section is essentially describing ways to reduce their effect on the final output. The important idea is the partition of the population into groups, each with its own fixed population of categories that must be conserved. Naturally, these groups are dynamic entities that split and merge according to the flow of play: the system will need to account for the flux of these conserved quantities as they move from group to group. The process is illustrated in Figure 8.6. We assume that the model is correct at frame k_0 – the start of the sequence. In frame k_1 the two smaller groups are merged, and so the conserved quantity associated with this group is simply the sum of the quantities from the two smaller groups. In frame k_2 the group splits into two once more and so must the conserved quantities too. This is the point at which the model is vulnerable to the introduction of errors. The conserved quantities must be correctly distributed between the two subsequent groups, otherwise each group will thereafter be preserving an incorrect quantity.

The groups are defined by spatial proximity of the players in the ground plane co-ordinate system. Two targets, t_1 and t_2, are in the same group g_i if $|\hat{\mathbf{x}}_1 - \hat{\mathbf{x}}_2| < d_0$. This condition applies exhaustively throughout the set of targets: thus, t_1 and t_2 may also be in the same group if there is another target t_3, situated within a distance d_0 of both of them. (This definition of a group is not related to the 'group' used in the first stage of processing, where players are occluding one another from a particular camera viewpoint.) The value of d_0 is empirically chosen at around 3–5 m, with the intention of keeping any uncertainty about players' identity inside the groups rather than between them. This process results in the set of groups $\{g_i\}$ at frame k.

The next step is to compute the associations that exist between the set of groups at frame k and the set of groups at $k + 1$: there is a straightforward definition – a direct association exists if there is any target included by both $g(k)$ and $g(k + 1)$. Furthermore, the union of those groups with a mutual direct association defines the

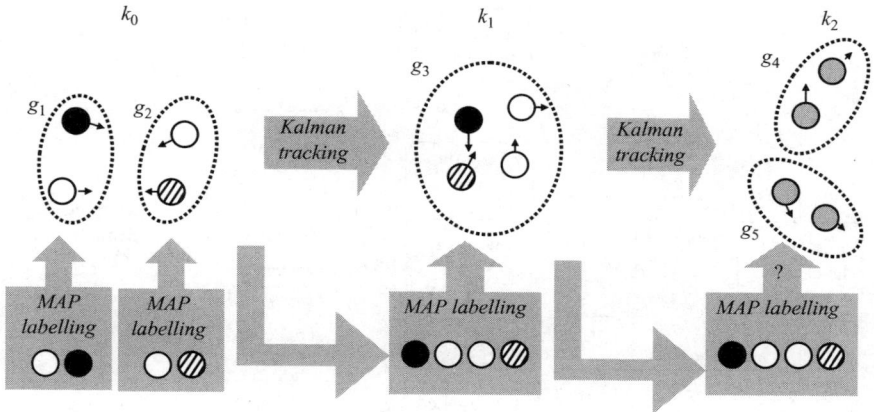

Figure 8.6   Diagrammatic example of target labelling, given local fixed population constraints. Here, two groups of players (g_1 and g_2) draw together to form a single group (g_3), which then splits into two groups (g_4 and g_5). Players in a group are labelled, subject to the local fixed population constraint, calculated from the previous frame. This is straightforward when groups merge (e.g. frame k_1) but less so when groups split (e.g. frame k_2)

set of targets involved in a split or merge event. It is through this union of groups that the fixed population is conserved: it is counted for frame k, and assigned for frame $k - 1$, using the joint MAP approach described in Sections 8.5.2 and 8.5.3.

The complete process cycle for the second stage is shown in Figure 8.7. First, the complete model ('labelled state') at frame k is subjected to the standard predict–associate–update steps described in Section 8.4, to obtain a 'pre-labelled' state for frame $k + 1$. It is called 'pre-labelled' because at this point the labels in the model are no longer trusted: errors may have been introduced, both in the data association stage and in the creation and deletion of tracks in the model. The next step is the re-labelling of this model, by finding the groups and associating them with the groups from the previous frame, as illustrated in Figure 8.7. These earlier groups determine the number of labels from each category that need to be assigned to the tracks in this most recent frame. This assignment is computed using the joint MAP method outlined in Section 8.5.2. This method requires as input, first the local fixed population constraint (the number of labels from each category) and second the model estimates (positional pdf and category probability distribution). Its output is a label for each track in the model.

This method has been implemented and tested on several sequences: details of the results are presented in Section 8.6. Although it preserves accuracy and coherence for longer periods than a purely global population constraint, it cannot recover from its mistakes. In the next section this problem is described, and possible solutions are discussed.

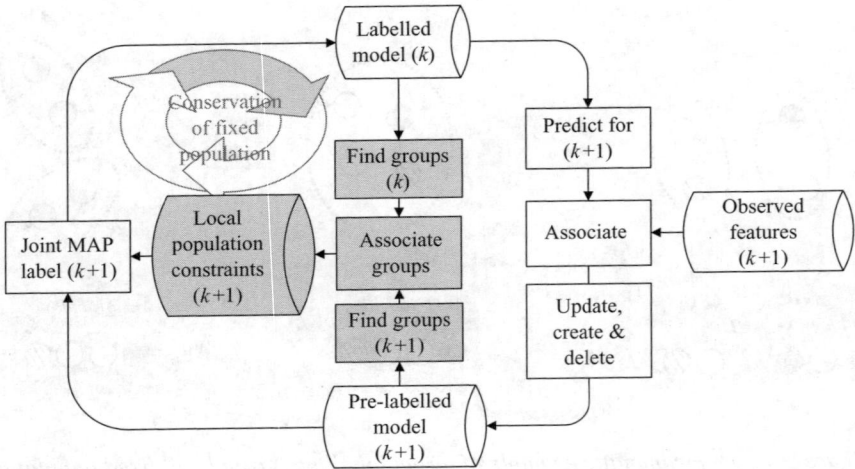

Figure 8.7   *Flow chart for the tracking process with local fixed population con-*
straints. Players are formed into groups, both at frame k and at k + 1.
These are associated with each other, to form the local population con-
straint used to make the MAP label estimate for each group. The fixed
population is conserved through the groups, rather than the individual
targets, since the categories of the latter are more difficult to estimate
when they are grouped together

8.5.5   Techniques for error detection and correction

For any system tracking multiple targets, occlusions and other interactions between
targets present the principal difficulty and source of error. Even though errors are
introduced when players come together to form a group, they only become apparent
when the players move apart and split this group.

This is reflected in the framework introduced above. When two groups merge,
the resulting population constraint is easily calculated as the sum of the constraints
for the two groups. When a group splits into two, there is frequently more than one
way that the population constraint (for each category) can be split – see the example
in Figure 8.6. The properties of the groups, and the tracks they include, are used
to estimate the most likely split. In principle, all possible combinations could be
evaluated; in our current system, we only explore the sub-space of combinations that
are consistent with the number of tracks allocated to each track.

The example in Figure 8.6 shows a group of four players, modelled as having split
into two groups, each of two players. We find which one of the six different ways
the labels can be associated with the tracks has the greatest likelihood of generating
the track properties we observe. However, it is also possible that the group split into
groups of three and one (or one and three), rather than the two groups of two modelled
by the tracking system. At present we do not consider these alternatives.

8.6 Results

The two-stage method outlined in this chapter has been tested in several recorded matches. A live system is being installed and will be available for testing soon. The advantages of multi-view tracking over single-view tracking can be demonstrated using the following two examples.

Figure 8.8 is an example to show the improved visibility by using multi-view tracking. In Figure 8.8(a), captured by camera C3, three players (A, B and C) are grouped together and present as a single foreground region. The foot position measurement of this foreground region basically reflects that of player A, who is the closest to the camera. Therefore, the other two players are missing in the measurements of camera C3 (see Figure 8.8(b)). However, these three players are well separated in Figure 8.8(c), captured by camera C1. Therefore, all three players are represented in the overall measurements (see Figure 8.8(e)), where player A is jointly detected

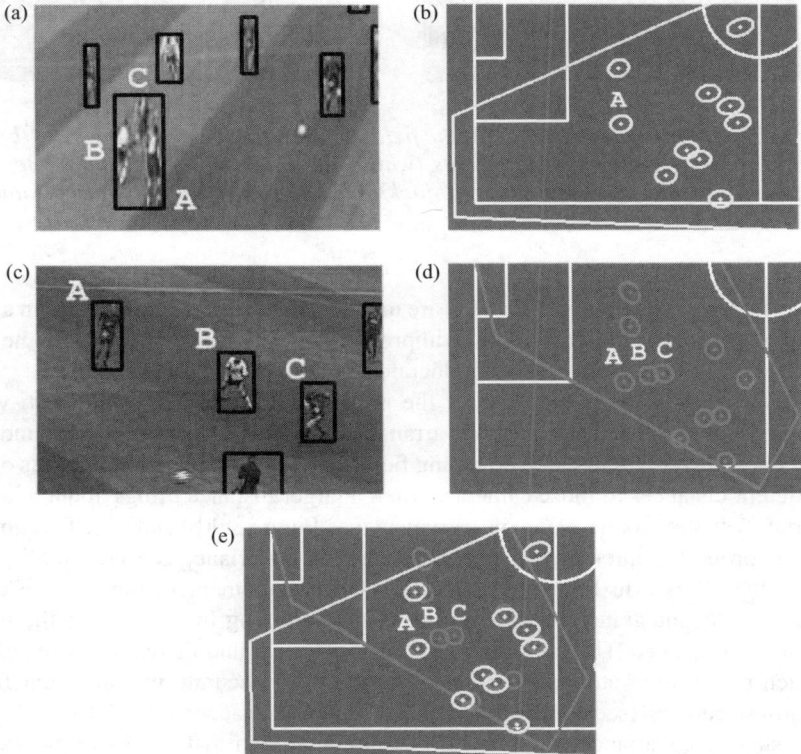

Figure 8.8   Improved visibility for a group of players: (a), (b) measurements of camera C3; (c), (d) measurements of camera C1; and (e) the overall measurements

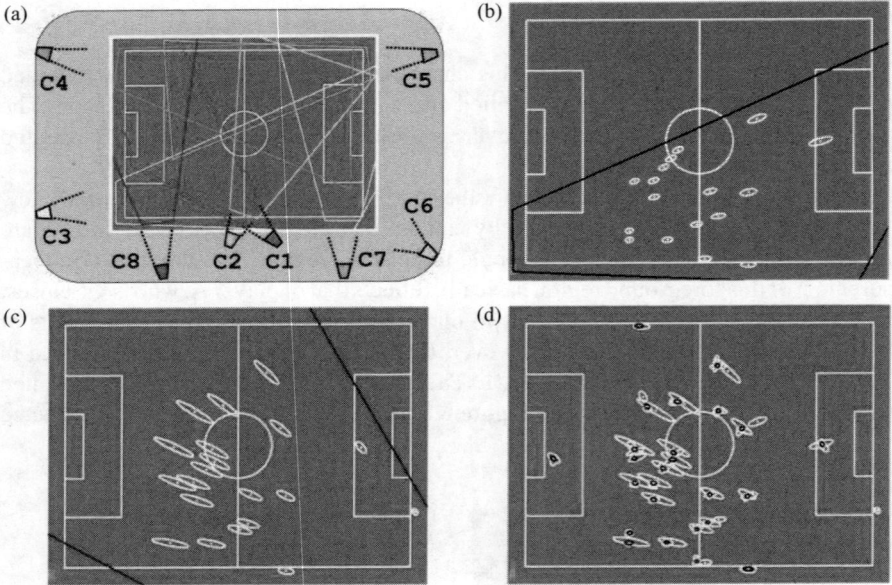

Figure 8.9   Improved accuracy: (a) the fields-of-view for the eight cameras; (b), (c) single-view measurements (white) projected on the ground plane; and (d) fused measurements (black). The ellipses represent the covariance

by two cameras while players B and C are uniquely detected by one camera. In addition to overlapping fields-of-view, the improved visibility is also reflected in the full coverage of the whole pitch with a moderate resolution (see Figure 8.9(a)).

Figure 8.9 is an example to show the improved accuracy on using multi-view tracking. Figure 8.9(a) illustrates our arrangement of the eight cameras around the pitch. Each player within the overlapping fields-of-view of multiple cameras is often at different distances to those cameras. Often a larger distance brings about a larger uncertainty in the ground-plane measurement (see Figure 8.9(b) and (c)). This uncertainty is properly represented by the measurement covariance $\mathbf{R}$ (indicated by the size of the ellipses in Figure 8.9(b)–(d)). The measurement fusion scheme used in this chapter integrates the measurements by weighting the inverse of the measurement covariance. Therefore, the fused measurement and thereafter the estimate for each player are automatically biased to the most accurate measurement from the closest camera (see Figure 8.9(d)). The improved accuracy is also reflected in the smaller covariance of the fused measurement than that from any individual camera.

Figure 8.10 shows the trajectories of the tracked players over a long time period. The non-linear trajectories (except those of two linesmen) not only reflect the manoeuvres of these players but also indicate that most of these players are tracked across

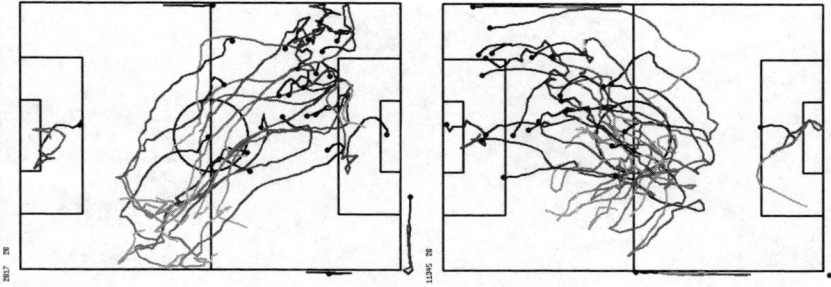

Figure 8.10 The trajectories of players over a long time. Grey scale indicates the elapsed time

Table 8.1 Multi-view tracking evaluation with target numbers. The ground-truth target number is 25 in both cases

Sequences	Newcastle vs Fulham	Arsenal vs Fulham
Number of tested frames	5000	5000
Mean, μ, of target number	26.63	26.12
Standard deviation, σ, of target number	1.55	1.56
90% confidence interval of target number ($\mu \pm 1.65\sigma$)	26.63 ± 2.56	26.12 ± 2.58
95% confidence interval of target number ($\mu \pm 1.96\sigma$)	26.63 ± 3.04	26.12 ± 3.06

multiple camera fields-of-view. For any target with a missing measurement, our tracking algorithm uses its linear prediction as the estimate and thus generates a linear trajectory.

To evaluate the performance of our multiview tracking algorithm, we counted the number of tracked targets before the target selection step. There are several reasons for this: (1) it is inconvenient and laboured to extract the ground truth in a multi-camera environment with crowded targets; (2) when some target is undetected (most likely packed with another target) for a long time, it is automatically terminated and the target number decreases; (3) often a tracking error (a wrong data association) causes a 'new' target being established and thus the target number increases. We evaluated our algorithm with two sequences of 5000 frames and the results are shown in Table 8.1. It is not surprising that the mean of target number is slightly larger than 25 (20 outfield players, 2 goalkeepers and 3 referees), because our algorithm encourages those targets with missing measurement to be maintained for some time. It is also noted that the relative tracking error rates ($1.65\sigma/\mu$) are low enough (within ± 10 per cent) in 90 per cent of the tested frames. The remaining 10 per cent of the tested frames with larger tracking errors corresponds to some over-crowded situations, where players

Figure 8.11   Multi-view and single-view tracking at one frame. The surrounding images (from top-left to top-right) correspond to cameras C4, C3, C8, C2, C1, C7, C6 and C5

are tightly packed in pairs, for example, during a corner kick, or where camera calibration errors and measurement errors are compatible to the small distance between comparable players.

This initial tracking can be further refined in the target selection step, with the domain knowledge and the sub-optimal search over the player likelihood applied. An example of the multi-view tracking after target selection, along with the single-view detection and tracking at the same frame, is demonstrated in Figure 8.11.

8.7   Conclusions

An application of football player tracking with multiple cameras has been presented. The overall visibility of all the players is an advantage of the underlying system over existing single-view systems. The dynamic occlusion problem has been greatly relieved by using our single view tracking with partial observations and multi-sensor tracking architecture. The estimate of each player is automatically biased to the most accurate measurement of the closest camera, and the fused measurement in overlapping regions is more accurate than that from any individual camera. The undergoing work includes using probabilistic data association methods, such as multiple hypothesis tracking and JPDAF [13], to improve the ground-plane tracking, and feeding back the multi-view tracking result to the single-view tracker, which further improves the multi-view tracking.

Acknowledgements

This work is part of the INMOVE project, supported by the European Commission IST 2001-37422. The project partners include Technical Research Centre of Finland, Oy Radiolinja Ab, Mirasys Ltd, Netherlands Organization for Applied Scientific Research TNO, University of Genova, Kingston University, IPM Management A/S and ON-AIR A/S. We would like to thank Dr James Black at Kingston University for his help in computing the Jacobian matrix of coordinate transformation.

References

1 S.S. Intille and A.F. Bobick. Closed-world tracking. In *Proceedings of ICCV*, 1995, pp. 672–678.
2 Y. Seo, S. Choi, H. Kim, and K.S. Hong. Where are the ball and players?: soccer game analysis with color-based tracking and image mosaic. In *Proceedings of ICIAP*, 1997, pp. 196–203.
3 A. Yamada, Y. Shirai, and J. Miura. Tracking players and a ball in video image sequence and estimating camera parameters for 3D interpretation of soccer games. *Proceedings of ICPR*, 2002, pp. 303–306.
4 T. Bebie and H. Bieri. SoccerMan: reconstructing soccer games from video sequences. *Proceedings of ICIP*, 1998, pp. 898–902.
5 C. Needham and R. Boyle. Tracking multiple sports players through occlusion, congestion and scale. *Proceedings of BMVC*, Manchester, 2001, pp. 93–102.
6 Q. Cai and J.K. Aggarwal. Tracking human motion using multiple cameras. In *Proceedings of ICPR*, 1996, pp. 68–72.
7 S. Khan, G. Javed, Z. Rasheed, and M. Shah. Human tracking in multiple cameras. In *ICCV'2001*, Vancouver, 2001, pp. 331–336.
8 G. Stein. Tracking from multiple view points: self-calibration of space and time. In *Proceedings of DARPA IU Workshop*, 1998, pp. 1037–1042.
9 J. Black, T. Ellis, and P. Rosin. Multi-view image surveillance and tracking. *Proceedings of IEEE Workshop on Motion and Video Computing*, Orlando, 2002, pp. 169–174.
10 R. Collins, A. Lipton, H. Fujiyoshi, and T. Kanade. Algorithms for cooperative multisensor surveillance. *Proceedings of the IEEE*, 2001;89(10):1456–1477.
11 R. Tsai. An efficient and accurate camera calibration technique for 3D machine vision. In *Proceedings of CVPR*, 1986, pp. 323–344.
12 J. Ren, J. Orwell, G. Jones, and M. Xu. A general framework for 3D soccer ball estimation and tracking. In *Proceedings of ICIP*, Singapore, 2004, pp. 1935–1938.
13 Y. Bar-Shalom and X.R. Li. *Multitarget–Multisensor Tracking: Principles and Techniques*. YBS, 1995.
14 C. Stauffer and W.E. Grimson. Adaptive background mixture models for real-time tracking. In *Proceedings of the CVPR*, 1999, pp. 246–252.

15 M. Xu and T. Ellis. Partial observation vs. blind tracking through occlusion. In *Proceedings of BMVC*, 2002, pp. 777–786.

16 M.J. Swain and D.H. Ballard. Colour indexing. *International Journal of Computer Vision*, 1991;7(1):11–32.

17 G. Jaffre and A. Crouzil. Non-rigid object localization from color model using mean shift. In *Proceedings of ICIP*, Barcelona, Spain, 2003, pp. 317–320.

18 T. Kawashima, K. Yoshino, and Y. Aoki. Qualitative image analysis of group behaviour. In *Proceedings of CVPR*, 1994, pp. 690–693.

Chapter 9

A hierarchical multi-sensor framework for event detection in wide environments

G. L. Foresti, C. Micheloni, L. Snidaro and C. Piciarelli

9.1 Introduction

Nowadays, the surveillance of important areas such as shopping malls, stations, airports and seaports, is becoming of great interest for security purposes. In the last few years, we have witnessed the installation of a large number of cameras that now are almost covering every place in the major cities. Unfortunately, the majority of such sensors belong to the so-called second generation of CCTV surveillance systems where human operators are required to interpret the events occurring in the scene. Only recently modern and autonomous surveillance systems have been proposed to address the security problems of such areas. Many systems have been proposed [1–6] for a wide range of security purposes from traffic monitoring [7,8] to human activity understanding [9]. Video surveillance applications often imply paying attention to a wide area, so different kinds of cameras are generally used, for example, fixed cameras [3,5,6], omni-directional cameras [10,11] and pan, tilt and zoom (PTZ) cameras [4,12,13].

The use of these kinds of cameras requires that their number and placement must be fixed in advance to ensure an adequate monitoring coverage of the area of interest. Moreover, developed systems usually present a logical architecture where the information acquired by the sensors is processed by a single operative unit.

In this chapter, we propose a hierarchical framework representing a network of cameras organised in sub-nets in which static and moving (PTZ) sensors are employed. The logical architecture has been designed to monitor motion inside a local area by means of static networks. Furthermore, the PTZ sensors can be tasked, as a consequence of an event detection, to acquire selected targets with higher resolution. Such a property gives an augmented vision to the system that allows better understanding of the behaviours of the objects of interest.

Information about the trajectories of the moving objects, computed by different sensors, is sent, bottom-up, through the hierarchy to be analysed by higher modules. At the first level, trajectories are locally fused to detect suspect behaviours inside local areas. Such trajectories are sent to higher nodes that analyse the activities that have occurred through the different local areas. This allows the detection of events with higher complexity.

The novelty of the proposed system resides in a reliable method of computing the trajectory of the objects using data from different sensors, in the use of an autonomous active vision system for the target tracking and in a scalable system for the detection of the events as their complexity increases.

9.2   System description

The proposed system is composed of a network of cooperating cameras that give a solution to the problem of wide area monitoring (e.g. parking lots, shopping malls, airport lounges). Specifically, the system could employ a generic number of both static and active sensors in a scalable and hierarchic architecture where two main types of sub-systems have been considered: (1) static camera system (SCS) and (2) active camera system (ACS).

As shown in Figure 9.1, the hierarchy of the system is given by two main entities: the local area monitoring networks (LAMNs) and the operative control unit (OCU).

LAMNs are organised in a hierarchical way to provide the monitoring of a local area. Such an architecture employs the SCSs to detect and track all the objects moving within the assigned area, and the ACSs.

The placement is done balancing two opposite needs: on the one hand the possibility to cover the entire area minimising the number of sensors, while on the other the need for redundancy to counteract targets occlusions by providing overlapping fields of view. First level nodes compute trajectories and send important features to the OCU, the highest level node (root node) of the hierarchy. Then, all the trajectories computed by the first level nodes are analysed to detect possible dangerous actions.

In that case, trajectory data are used to task one or more ACSs to autonomously track the suspect object with higher definition. ACSs do not require a continuous setting of the sensor position since they are able to track and to maintain the object of interest inside their field of view. Nevertheless, we studied a policy which allows checking the correctness of the ACS tracking. This process is performed in three steps: (1) estimating the rotation angles needed to get the ACS on the target (i.e. the pan and tilt rotations to get the target in the centre of the field of view); (2) requesting the current state (the rotation angles) of the active camera; and (3) checking if the current ACS state is coherent with the target state (2D coordinates in the top-view map).

At the same level of the LAMNs, a communication system has been provided to deliver information of interest throughout different networks. It is interesting to note

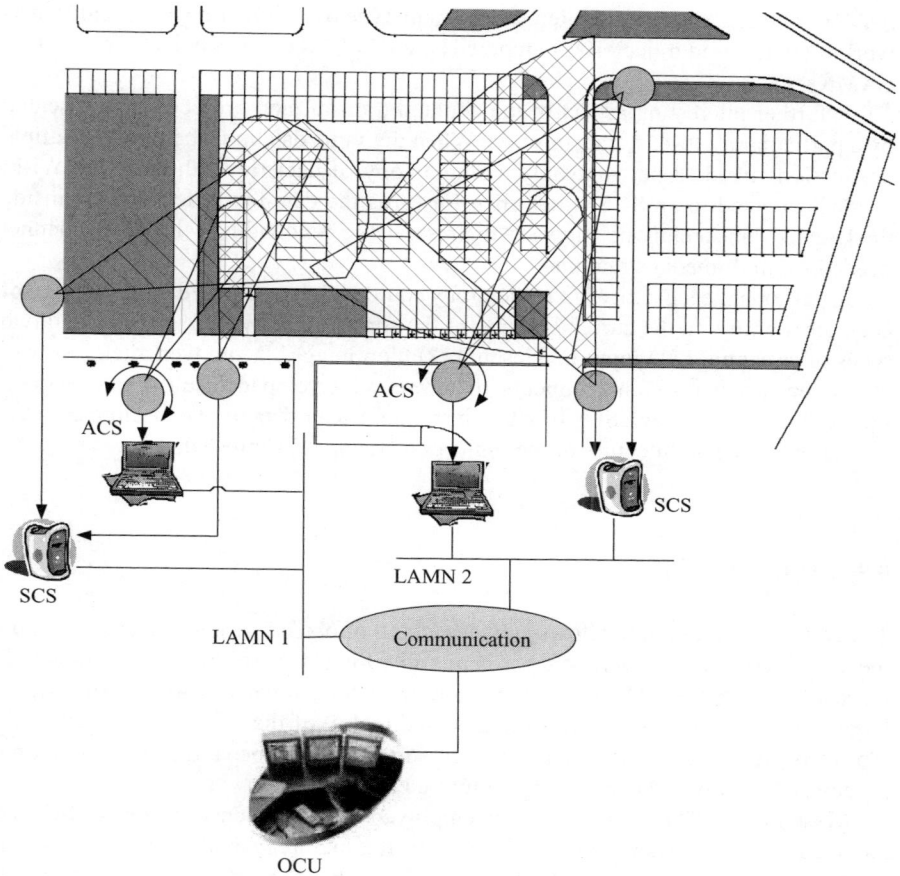

Figure 9.1   System architecture

that since the monitored area could be very extensive a complete wiring of the net-works was not available. To overcome such a problem, the developed communication protocol can be used with wireless networks and with multi-cast properties that allow reductions in the required bandwidth for the communication.

At the highest level of the hierarchy we developed a centralised control unit, the OCU, where all the information collected by the single networks is received and processed to understand the activities occurring inside the monitored area.

Finally, a custom-built communication protocol allows transporting such infor-mation from the LAMNs to the OCU. It is composed of two main components: (1) Wi-Fi communication among different LAMNs and (2) a software algorithm to determine both the data to be sent and the destinations of the messages. The protocol allows each SCS system inside a network to request data computed by other networks. For example, an SCS could request an ACS system inside another network to track an object of interest. The main type of information sent inside a network is about fused trajectories computed by either SCS or ACS trajectories in other LAMNs.

A further interesting aspect of the adopted transmission protocol is represented by a particular multi-cast transmission. When the same data must be sent to multiple destinations inside a network only a single copy of data flows through the Wi-Fi channel directed to an SCS which then forwards the data to all destinations inside the LAMN. This technique allows us to preserve the bandwidth of the Wi-Fi channel needed for multimedia streaming.

In particular, each entity of the hierarchy could request multimedia information of each other entity. In the current version of the surveillance system, we included two types of streaming: (1) input images and (2) blob images. Moreover, such a stream can be sent also to different recipients; therefore, we developed a multi-cast streaming protocol that allows each SCS or ACS to stream proper data to one or more entities. In Figure 9.2 the architecture of the multi-cast streaming is presented.

9.3   Active tracking

The active camera system allows active tracking of objects of interest. Precisely, by means of active vision techniques it is possible to customise the image acquisition for security purposes. Moreover, the capacity to control gaze as well as to acquire targets at high resolution gives an augmented reality of the monitored scene. Such a property turns out to be fundamental to supporting the behaviour understanding process, especially if the area to be monitored is really wide.

Most times, PTZ cameras when employed in surveillance systems [14] are included as slave systems inside a master–slave hierarchy with static cameras as masters. Such cameras, therefore, continuously control the gaze of the PTZ cameras by computing position parameters. Instead, in the proposed solution, the ACSs are entities able to track autonomously a selected target.

The logical architecture shown in Figure 9.3 shows the modules developed to allow the cooperation of an active sensor with the network of static sensors. In particular, a communication system can deliver a request for focusing the attention of the active sensor on a selected target inside the monitored area. Such a request is fulfilled by changing the pan and tilt parameters of the PTZ unit to head the camera towards the spot occupied by the target.

When the repositioning is accomplished, the tracking phase starts. To track the objects either while panning and/or tilting or while zooming, two different tracking techniques have been considered and developed. However, at the first iteration, in order to detect the object of interest, an image alignment computation is required. In successive frames, to acquire the target with the highest resolution, the system is able to decide if a zoom operation is required or not. In the first case, a set of features are extracted from the object and tracked. Then the singular displacements are used to compute a fixation point needed to determine the pan, tilt and zoom parameters for the

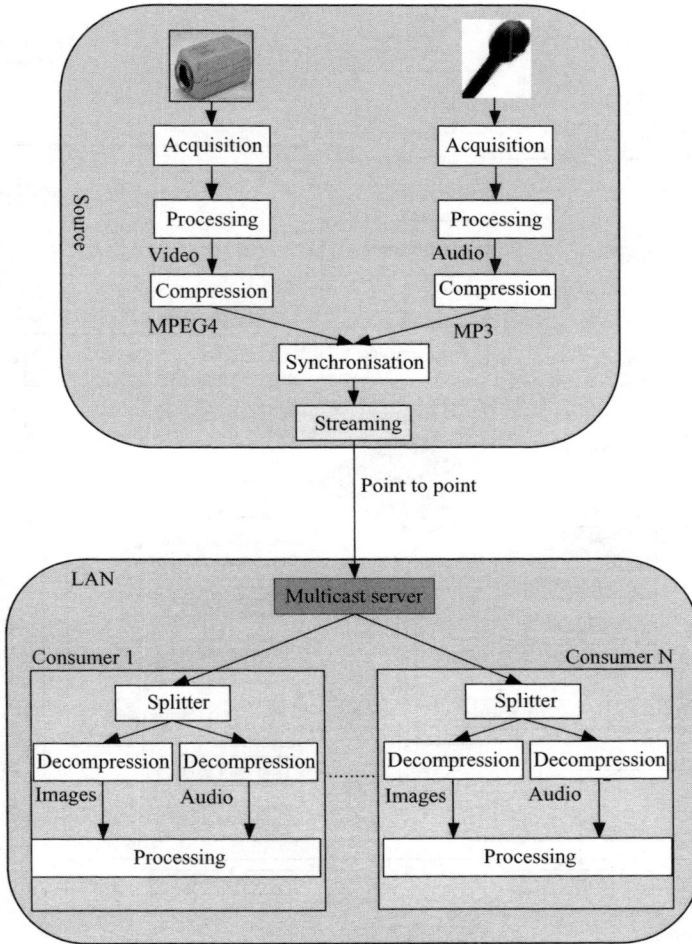

Figure 9.2   Logic architecture of the developed multi-cast stream system. A multi-cast server receives a single stream that is redirected to all the recipients inside their own network. In our case the multi-cast server is a service run on an SCS

next time instant. When a zoom operation is not requested, a registration technique based on the computation of an affine transform is executed.

Finally, as a consequence of the object detection method adopted, the tracking phase uses the meanshift technique [15] when the blob is detected and a Kalman filter technique [16] when the previous step has carried out the fixation point.

In the first case, a frame by frame motion detection system [13] is used. In such a system, a major speed-up technique consists in the use of a translational model to describe the transformation between two consecutive frames. This allows

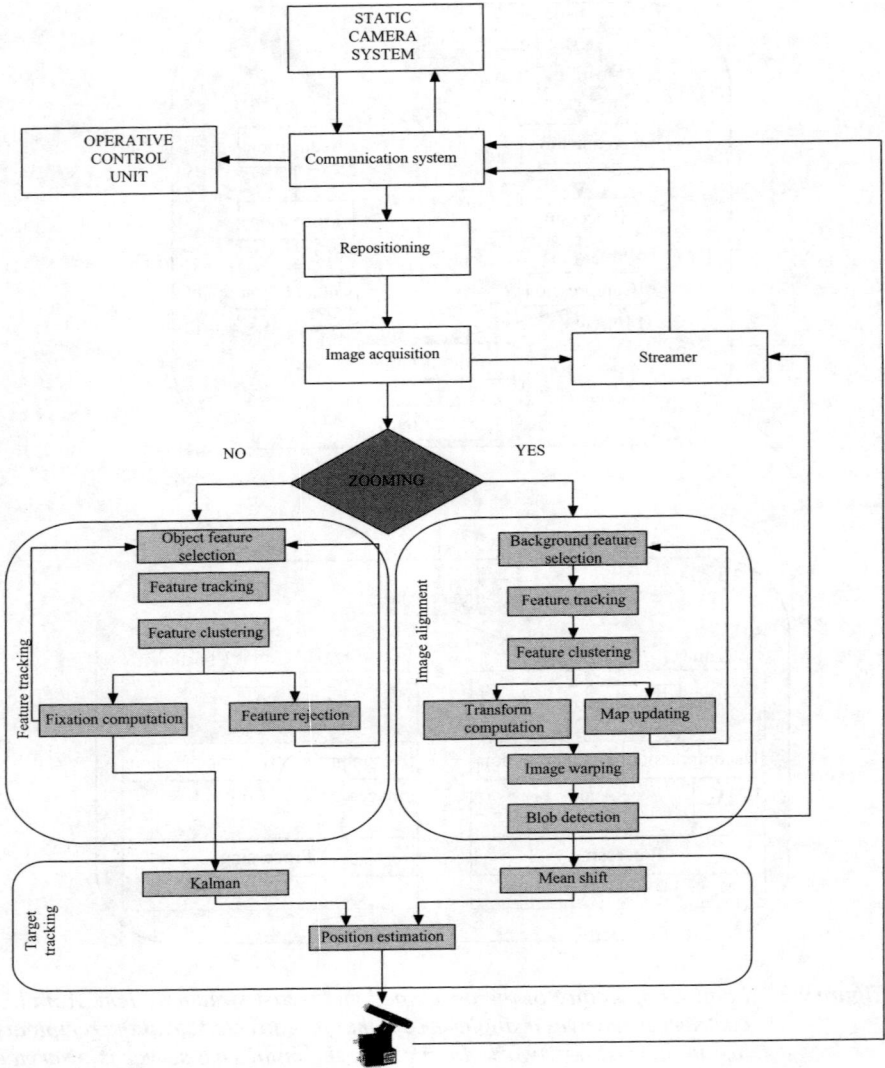

Figure 9.3   General architecture of the active camera systems. Two different approaches to detecting objects have been studied depending on whether a zoom operation is performed or not. As a consequence, two different object tracking methods have been taken into account

the computation of a simple displacement vector **d** between two consecutive frames i and $i + 1$ to compute the registration.

To compute the displacement vector **d**, a low number of features $f_j, j = 1, \ldots, 20$, representing a set of good trackable features TFS_i belonging to static objects, is selected and tracked by a feature tracking algorithm T [18]. Hence, the displacement estimation takes as input the list of local displacements $D_i = \{d_j | f_j \in T(\text{TFS}_i)\}$ and uses a feature clustering approach to estimate the displacement vector of the background $\mathbf{d}$. The image warping operation translates the current frame by the estimated displacement vector $\mathbf{d}$. A change detection operation [17] is applied between two consecutive frames $I(\mathbf{x}, i)$ and $I(\mathbf{x} + \mathbf{d}_{i+1}, i + 1)$ where the latter is the current frame after compensation.

The proposed image alignment method is based on the tracking algorithm proposed by Lucas *et al.* [18,19]. Thereafter, the feature tracking algorithm [13] is applied on it. Unfortunately, it often occurs that some features in the TFS are not tracked well due to noise, occlusions, etc. In order to deal with this problem, it is necessary to distinguish features tracked well from the others. Let $\mathbf{d}_j$ be the displacement of the feature f_j; we define $C(\mathbf{d}_c) = \{f_j \in \text{TFS}_i | \mathbf{d}_j = \mathbf{d}_c\}$ the cluster of all features having the displacement equal to the vector $\mathbf{d}_c$. Let $C = \{C(\mathbf{d}_c)\}$ be the set of all clusters generated by the tracking algorithm. For each of them a reliability factor RF is defined as an indicator of the closeness of its displacement to the background one. Let RF be defined as follows:

$$\text{RF}(C(\mathbf{d}_c)) = \frac{\sum_{f_j \in C(\mathbf{d}_c)} E_{f_j}}{|C(\mathbf{d}_c)|^2}, \tag{9.1}$$

where $E_{f_j} = \sum_{\mathbf{x} \in f_j} \sqrt{(f_j(\mathbf{x} + \mathbf{d}_{i+1}) - f_j(\mathbf{x}))^2}$ is the residual of the feature between its position in the previous and current frames and $|\cdot|$ represents the cardinality operator. According to the definition of the RF factor and the definition of the displacement clustering, the displacement of the cluster with the minimum RF factor is taken into account for the estimation of the vector $\mathbf{d}$.

Such a method has been applied on sub-areas of the image to also take into account rotation movements and the tracking with a higher zoom level. In particular, the image has been divided into nine areas in which the computation of displacement is performed. Finally, the three clusters with the best reliability factor are used to compute an affine transform.

When a zoom operation is involved, since the registration techniques are not reliable for detecting moving objects, we have adopted a technique that tracks a target by computing a fixation point from a set of features selected and tracked on the target. In particular, the Shi–Tomasi–Kanade method is used to track a set of features, this time extracted from a window (fovea) centred on the object of interest. The extraction of new features is performed when the current number of well tracked features is less than a pre-defined threshold n_{thr} (since the algorithm needs at least three features for the computation of the affine transform, we decided to set $n_{\text{thr}} = 5$).

To detect the clusters we need first to compute the affine transform A for each tracked object. Such a computation is performed over all the features belonging to an object. Let TFS_{obj} be the set of features extracted from a window (i.e. fovea) centred on the object of interest. The clustering technique introduced for the computation of the displacement between two consecutive frames is also used for detecting clusters

Figure 9.4   Example of the proposed alignment method. For each of the nine windows, the detected clusters are shown by presenting their features with different colours. The arrows denote the three features considered for the computation of the affine transform

of features belonging to the target:

$$C_{obj}(\mathbf{d}) = \{f_j \in \text{TFS}_{obj} | \|\tilde{\mathbf{d}}_j - \tilde{\mathbf{d}}\|_2 \leq r_{tot}\}. \tag{9.2}$$

Once the computation of all the clusters is concluded, we can easily find the background cluster and therefore all the features that have been erroneously extracted from the background in the previous feature extraction step. In particular, after applying the affine transform on the features, if these features belong to the background then they should have a null or small displacement. Therefore, to determine the background clusters we have adopted the following rule:

$$C_{obj}(\tilde{\mathbf{d}}_k) \in \begin{cases} \text{background} & \text{if } \|\tilde{\mathbf{d}}_k\|_2 \leq r_{tol} \\ \text{object} & \text{otherwise.} \end{cases} \tag{9.3}$$

From the resulting clusters we do not consider all those whose cardinality (i.e. number of features) is less than 3. This is due to the Tordoff–Murray [20] technique that we use to track the fixation point and which requires the computation of an affine transform for which we need at least three features.

After having deleted all the features that either belong to the background or to a cluster with cardinality lower than 3, we can apply the technique proposed by Tordoff–Murray [20] over each cluster to determine the fixation point. Let $\mathbf{g}'_l$ be the fixation point we need to compute for each object *l*; then we need to solve the following equation:

$$\mathbf{g}'_{\tilde{\mathbf{d}}_k} = N\mathbf{g} + \mathbf{r}, \tag{9.4}$$

where $\mathbf{g}'_{\tilde{\mathbf{d}}_k}$ is the new position of the fixation point for the set of features belonging to the cluster $C_{obj}(\tilde{\mathbf{d}}_k)$, computed from the position of the old fixation point $\mathbf{g}$ of the same object.

Figure 9.5   Example of active tracking during a zoom phase. Boxes represent tracked features organised in clusters while the dot represents the fixation point computed over the features belonging to the selected cluster

Following this heuristic, a fixation point is computed for each detected cluster allowing the estimation of the centre of mass of each moving object, thus avoiding the selection of features in the background or belonging to objects that are not of interest.

The ability of the proposed method to interleave the two developed techniques allows continuous and autonomous tracking of a selected target by maintaining it at a constant and predetermined resolution.

9.4   Static camera networks

The static camera networks are a set of static camera systems whose sensors have been installed in such a way as to monitor sub-areas of the environment. Each SCS is organised so that it can have a generic number of sensors for the monitoring of the assigned area. In particular, it receives input data coming from different static sensors that have a first coarse object tracking module to identify and track all the objects in the field of view. At the sensor level, object features are used to address the problem of the temporal correspondence of multiple objects. As a result, the position on a top-view map of the scene is also computed and sent to the SCS. Data computed by all the sensors connected to the SCS are processed by a first data fusion technique that allows us to obtain more robust information about the state of each object (position, velocity, etc.).

At the LAMN level, a different and more reliable object tracking technique, based on the analysis of the state vectors generated and shared by each SCS, has been adopted to track objects inside the entire local area. Hence, such tracking allows detection and classification of the occurring events. In the following sections, the techniques of tracking objects and classifying their actions inside the LAMNs are presented.

9.4.1   Target tracking

The system needs to maintain tracks for all the objects that exist in the scene simultaneously. Hence, this is a typical multi-sensor multi-target tracking

Figure 9.6   Example of continuous tracking by adopting the two techniques for object identification at different time instants. In particular, in the first row the image alignment technique is used to segment the object from the background. In the second row the fixation point on the target is computed during a zoom-out operation. Finally, new object identification through image alignment is done

problem: measurements have to be correctly assigned to their associated target tracks and target-associated measurements from different sensors have to be fused to obtain a better estimation of the target state.

A first tracking process may occur locally to each image plane. For each sensor, the system can perform an association algorithm to match the current detected blobs with those extracted in the previous frame. A number of techniques are available, spanning from template matching to features matching [14], to more sophisticated approaches [15].

Generally, a 2D top-view map of the monitored environment is taken as a common co-ordinate system [5,21], but even GPS may be employed to globally pinpoint the targets [14]. The first approach is obviously more straightforward to implement, as a well-known result from projective geometry states that the correspondence between an image pixel and a planar surface is given by a planar homography [22,23].

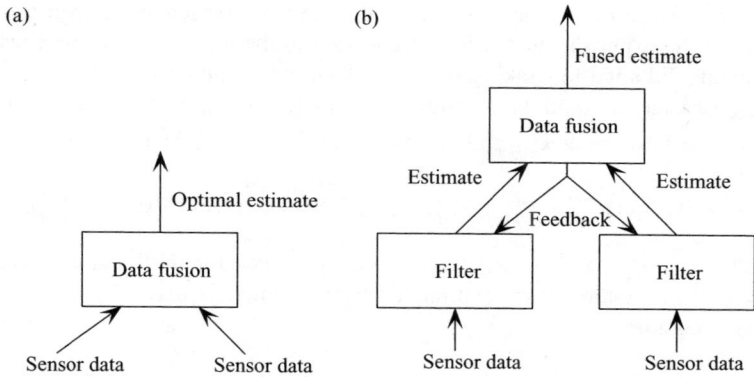

Figure 9.7 *(a) Optimal fusion and (b) track fusion*

The pixel usually chosen to represent a blob and be transformed into map coordinates is the projection of the blob centroid on the lower side of the bounding box [5,14,21].

To deal with the multi-target data assignment problem, especially in the presence of persistent interference, there are many matching algorithms available in the literature: nearest neighbour (NN), joint probabilistic data association (JPDA), multiple hypothesis tracking (MHT) and S–D assignment. The choice depends on the particular application, and detailed descriptions and examples can be found in [24–26], while recent developments in the subject may be found in [14,27–29].

9.4.2 Position fusion

Data obtained from the different sensors (extracted features) can be combined together to yield a better estimate. A typical feature to be fused is the target position on the map. A simple measurement fusion approach using Kalman filters [16,30] was employed for the purpose. This fusion scheme involves the fusion of the positions of the target (according to the different sensors) obtained right out of the coordinate conversion function, as can be seen in Figure 9.7(a). This algorithm is optimal, while the track-to-track fusion scheme, depicted in Figure 9.7(b), has less computational requirements but is sub-optimal in nature since fusing tracks (i.e. estimated state vectors) involves the decorrelation of the local estimates as they are affected by a common process noise and cannot be considered to be independent [24].

The measurement fusion not only is optimal in theory, but shows good results in practice as also reported in Reference 33. However, when dealing with extremely noisy sensors (i.e. video sensors performing poorly due to low illumination conditions), the track-to-track scheme is generally preferred. Running a Kalman filter for each track to obtain a filtered estimate of the target's position allows the smoothing of high variations due to segmentation errors. The actual scheme employed was a track-to-track without feedback, the rationale given by computational constraints. In fact, during the experiments, the high frequency of the measurements and real-time requirements did not allow taking feedback information into account.

The fused state $\hat{\mathbf{x}}_{k|k}$ on the position and velocity of a target is therefore obtained fusing the local estimates $\hat{\mathbf{x}}^i_{k|k}$ and $\hat{\mathbf{x}}^j_{k|k}$ from sensors i and j, respectively, as follows,

$$\hat{\mathbf{x}}_{k|k} = \hat{\mathbf{x}}^i_{k|k} + [\mathbf{P}^i_{k|k} - \mathbf{P}^{ij}_{k|k}][\mathbf{P}^i_{k|k} + \mathbf{P}^j_{k|k} - \mathbf{P}^{ij}_{k|k} - \mathbf{P}^{ji}_{k|k}]^{-1}(\hat{\mathbf{x}}^j_{k|k} - \hat{\mathbf{x}}^i_{k|k}), \quad (9.5)$$

where $\mathbf{P}^i_{k|k}$ and $\mathbf{P}^j_{k|k}$ are the error covariance matrices for the local estimates and $\mathbf{P}^{ij}_{k|k} = (\mathbf{P}^{ji}_{k|k})^T$ is the cross-covariance matrix, which is given by the following recursive equation:

$$\mathbf{P}^{ij}_{k|k} = [\mathbf{I} - \mathbf{K}^i_k \mathbf{H}^i_k][\mathbf{F}_{k-1}\mathbf{P}^{ij}_{k-1|k-1}\mathbf{F}^T_{k-1} + \Gamma_{k-1}\mathbf{Q}_{k-1}\Gamma^T_{k-1}][\mathbf{I} - \mathbf{K}^j_k \mathbf{H}^j_k], \quad (9.6)$$

where $\mathbf{K}^s_k$ is the Kalman filter gain matrix for sensor s at time k, Γ_k is the process noise matrix at time k, and $\mathbf{Q}$ is the process noise covariance matrix.

The process, for each target, involves the following steps: (1) collection of measurements available from the local sensors; (2) grouping and assignment of the measurements to each target known at the previous time instant; (3) updating each target's state by feeding the associated filtered estimates to the fusion algorithm.

The fusion procedure maintains its own list of targets. Note that the second step is performed with the constraint that only a single measurement from a given sensor is to be associated with a single target in the list maintained by the fusion procedure.

To regulate the fusion process automatically according to the performance of the sensors, a confidence measure is described next to weight local estimates.

The following measure, called appearance ratio (AR) [32], gives a value to the degree of confidence associated with the jth blob extracted at time t from the sensor s:

$$\text{AR}(B^s_{j,t}) = \frac{\sum_{x,y \in B^s_{j,t}} D(x,y)}{|B^s_{j,t}|C}, \quad (9.7)$$

where $D(x,y)$ is the difference map obtained as the absolute difference between the current image and the reference image, and C is a normalisation constant depending on the number of colour tones used in the image. The AR is thus a real number ranging from 0 to 1 that gives an estimation of the level of performance of each sensor for each extracted blob. In Figure 9.8 an experiment of sensor performance involving two cameras monitoring the same scene is shown. Two frames per sensor are shown in columns (a) and (c). Looking at the processed images of the top row of columns (b) and (d), we can see that the first camera (b) is giving good detection of the rightmost person (AR = 0.78), while the other two people have been merged in a single bounding box. The tracking algorithm correctly detected the merging of two objects and the AR value is therefore not computed, because the box is not referring to a real object. In column (d) we can appreciate the usefulness of multiple cameras in resolving occlusions as the three people are correctly detected and with high AR values. In the bottom row the reverse case is shown where the first camera helps

Figure 9.8   AR values of the blobs extracted by two sensors monitoring the same area

disambiguating the merging situation detected by the second camera. Furthermore, the second camera gives a better detection of the leftmost person.

AR values are then used to regulate the measurement error covariance matrix to weight position data in the fusion process. The following function for the position measurement error has been developed in Reference 32:

$$r(B_{j,t}^s) = \text{GD}^2(1 - \text{AR}(B_{j,t}^s)), \tag{9.8}$$

where GD is the gating distance [24]. The function is therefore used to adjust the measurement position error so that the map positions calculated for blobs with high AR values are trusted more (i.e. the measurement error of the position is close to zero), while blobs poorly detected (low AR value) are trusted less (i.e. the measurement error equals the gating distance).

9.4.3   Trajectory fusion

Trajectories computed by the LAMNs are sampled, modelled through cubic splines and sent to the OCU. To reduce bandwidth requirements, object positions are sampled every z time intervals and the cubic spline passing through the points $M(k)$ and $M(k+z)$ calculated [33].

Only the starting and ending points and tangent vectors are needed to define a spline segment. Each of them may span a pre-defined number of measurements (track points on the map) or, to improve bandwidth savings, a dynamically computed one. In particular, the sampling procedure may take into account whether the target is manoeuvring or not and regulate the sampling frequency accordingly (the trajectory of a manoeuvring target requires more spline segments to be described). The spline segments are then sent to the higher level nodes.

The trajectory fusion occurs at the OCU. This node is in charge of composing the trajectories of the targets by joining together the spline segments sent by the LAMNs. The joining itself poses no problem when the segments are provided by a single LAMN, as every spline segment is received along with the target ID so that the association is easily performed. When a target moves from a sub-area to another controlled by a different LAMN, an association algorithm is required to maintain the target ID.

The higher level nodes are the ideal processing elements for running trajectory analysis algorithms as they have a broader view of the behaviour of the targets. Depending on the specific application, the system can detect and signal to the operator dangerous situations (a pedestrian walking towards a forbidden area, a vehicle going in a zigzag or in circles, etc.) [34,35].

9.5   Event recognition

An event is characterised by a set of k classified objects over a sequence of n consecutive frames, $E(k,n)$, $k = 1,\ldots,K$ and $n = 1,\ldots,N$. The goal of the proposed system is to detect an event and classify it into a set of three levels of increasing dangerousness: (1) normal event, (2) suspicious event and (3) dangerous event. Suspicious and dangerous events generate two kinds of different alarm signals that are displayed on the man–machine interface of the remote operator. Two types of events have been considered: (1) simple events and (2) composite events.

In the considered environment, that is, a parking lot, a simple event is represented by a vehicle moving and/or stopping in allowed areas or a pedestrian walking with typical trajectories (e.g. almost rectilinear for a long number of frames). Suspicious events are represented by pedestrians walking with trajectories that are not always rectilinear, pedestrians stopping in a given parking area, pedestrians moving around

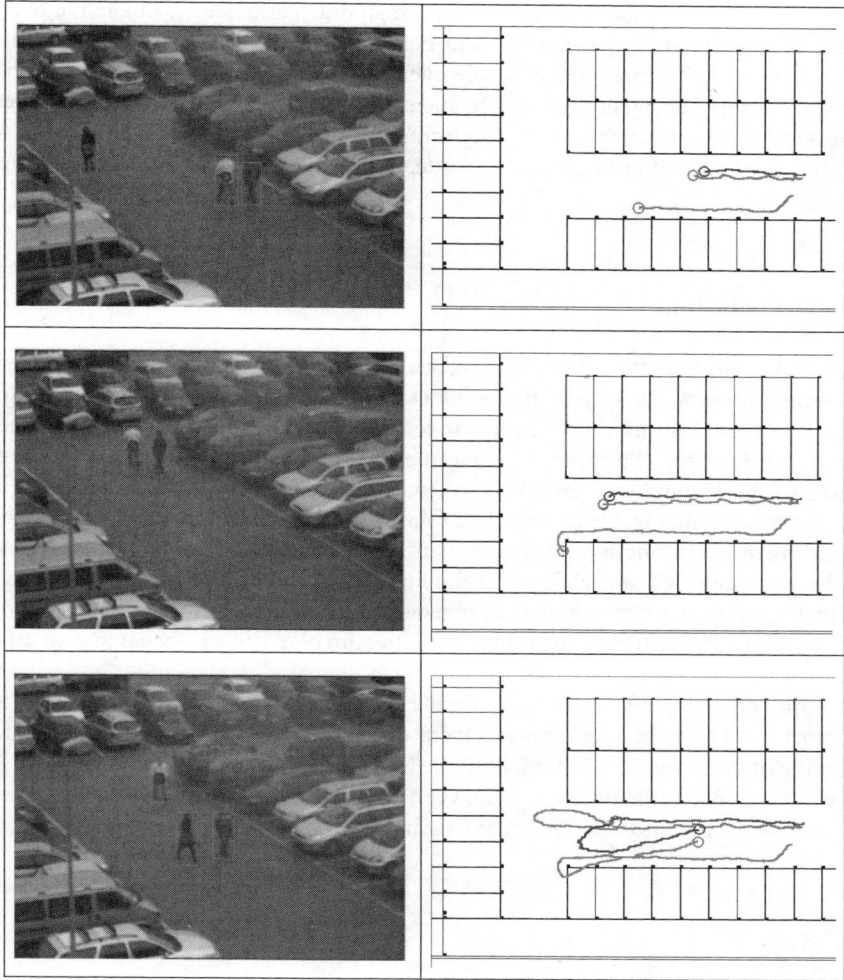

Figure 9.9   *Example of event recognition. Three people have been detected mov-*
ing inside the monitored environment. Their trajectories are consid-
ered atypical since all of them are going back and forth close to
the parked cars. In particular, one person (darker track) has been
considered to be actively tracked

a vehicle, etc. Dangerous events are represented by pedestrians or vehicles moving or stopping in forbidden areas, pedestrians moving with atypical trajectories
(e.g. moving around several vehicles as shown in Figure 9.9).

For each event, the initial and final instants, t_{in} and t_{fin}, the class C_i of the detected
object, the coefficients of the spline which approximates the object trajectory, the
average object speed, S_i, and the initial and final object position in the 2D top-view

map, $pos_i(x_{tin}, y_{tin})$ and $pos_i(x_{tfin}, y_{tfin})$, are stored. Periodically, the active events are analysed to verify if some of them bear some correlations. If there exist two or more temporally consecutive simple events with initial or final position spatially closed, a composite event is generated and stored. When a composite event has been detected, its classification is easily performed by analysing if it is contained in a set of models of normal or unusual composite events previously defined by the system operator.

9.6   Conclusions

In this chapter, a multi-sensor framework exploiting both static and active cameras has been proposed for solving the problem of monitoring wide environments. How cameras can be organised in networks to local monitor sub-areas of the environment has been presented. Furthermore, how the tracking can be performed at different levels of the hierarchy to give higher semantic information as the trajectories are processed by higher levels has also been shown. Then, the OCU analyses the actions occurring inside the monitored environment so as to detect dangerous events. As soon as these are detected at each level of the hierarchy, active cameras can be tasked to autonomously track the responsible objects. Moreover, recordings of such actions can be sent via multi-cast to different entities involved in the behaviour analysis process.

Further developments will look into the possibility of configuring the acquisition parameters such as the zoom, iris and focus of remote sensors to allow a better understanding of the scene. In addition, multiple hypothesis analysis will be considered for investigating better the interaction between different moving objects so as to be able to forecast dangerous events in order to activate proper counter-measures.

References

1 C. Regazzoni, V. Ramesh, and G. Foresti. Special issue on video communications, processing, and understanding for third generation surveillance systems. *Proceeding of the IEEE*, 2001;89(10):1355–1539.
2 L. Davis, R. Chellapa, Y. Yaccob, and Q. Zheng. Visual surveillance and monitoring of human and vehicular activity. In *Proceedings of DARPA97 Image Understanding Workshop*, 1997, pp. 19–27.
3 R. Howarthand and H. Buxton. Visual surveillance monitoring and watching. In *European Conference on Computer Vision*, 1996, pp. 321–334.
4 T. Kanade, R. Collins, A. Lipton, P.Burt, and L.Wixson. Advances in cooperative multisensor video surveillance. In *Proceedings of DARPA Image Understanding Workshop*, Vol. 1, November 1998, pp. 3–24.
5 G.L. Foresti. Object recognition and tracking for remote video surveillance. *IEEE Transaction on Circuits and Systems for Video Technology*, 1999;9(7):1045–1062.

6 G. Foresti, P. Mahonen, and C. Regazzoni. *Multimedia Video-Based Surveillance Systems: from User Requirements to Research Solutions.* Kluwer Academic Publisher, 2000.

7 D. Koller, K. Daniilidis, and H.H. Nagel. Model-based object tracking in monocular sequences of road traffic scenes. *International Journal of Computer Vision*, 1993;10:257–281.

8 Z. Zhu, G. Xu, B. Yang, D. Shi, and X. Lin. VISATRAM: a real-time vision system for automatic traffic monitoring. *Image and Vision Computing*, 2000;18(10): 781–794.

9 S. Dockstader and M. Tekalp. Multiple camera tracking of interacting and occluded human motion. *Proceedings of the IEEE*, 2001;89(10):1441–1455.

10 J. Gluckman and S. Nayar. Ego-motion and omnidirectional camera. In *IEEE International Conference on Computer Vision*, Bombay, India, January 3–5, 1998, pp. 999–1005.

11 S. Nayar and T. Boult. Omnidirectional vision systems. In *Proceedings of the DARPA Image Understanding Workshop*, New Orleans, May 1997, pp. 235–242.

12 D. Murray and A. Basu. Motion tracking with an active camera. *IEEE Transactions on Pattern Analysis and Machine Intelligence*, 1994;19(5): 449–454.

13 G. Foresti and C. Micheloni. A robust feature tracker for active surveillance of outdoor scenes. *Electronic Letters on Computer Vision and Image Analysis*, 2003;1(1):21–34.

14 R. Collins, A. Lipton, H. Fujiyoshi, and T. Kanade. Algorithms for cooperative multisensor surveillance. In *Proceedings of the IEEE*, 2001;89:1456–1477.

15 D. Comaniciu, V. Ramesh, and P. Meer. Real-time tracking of non-rigid objects using mean shift. In *Proceedings of the IEEE Conference on Computer Vision and Pattern Recognition*, Hilton Head, South Carolina, 2000, pp. 142–149.

16 D. Willner, C.B. Chang, and K.P. Dunn. Kalman filter algorithms for a multisensor system. *Proceedings of the IEEE Conference on Decision and Control*, 1978, pp. 570–574.

17 B. Lucas and T. Kanade. An iterative image registration technique with an application to stereo vision. In *Proceedings of the 7th International Joint Conference on Artificial Intelligence*, Vancouver, August 1981, pp. 674–679.

18 L. Snidaro and G.L. Foresti. Real-time thresholding with Euler numbers. *Pattern Recognition Letters*, 2003;24(9–10):1533–1544.

19 C. Tomasi and T. Kanade. *Detection and Tracking of Point Features.* Carnegie Mellon University, Pittsburgh PA, Tech. Rep. CMU-CS-91-132, 1991.

20 B.J. Tordoff and D.W. Murray. Reactive control of zoom while tracking using perspective and affine cameras. *IEEE Transactions on Pattern Analysis and Machine Intelligence*, 2004;26(1):98–112.

21 G.L. Foresti. Real-time detection of multiple moving objects in complex image sequences. *International Journal of Imaging Systems and Technology*, 1999;10:305–317.

22 R. Tsai. A versatile camera calibration technique for high-accuracy 3D machine vision metrology using off-the-shelf TV cameras and lenses. *IEEE Journal of Robotics and Automation*, 1987;RA3(4):323–344.

23 O.D. Faugeras, Q.-T. Luong, and S.J. Maybank. Camera self-calibration: theory and experiments. In *Proceedings of European Conference on Computer Vision*, 1992, pp. 321–334.

24 Y. Bar-Shalom and X. Li. *Multitarget–Multisensor Tracking: Principles and Techniques*. YBS Publishing, 1995.

25 S.S. Blackman. *Multiple-Target Tracking with Radar Applications*. Artech House, 1986.

26 A.B. Poore. Multi-dimensional assignment formulation of data association problems arising from multi-target and multi-sensor tracking. *Computational Optimization and Applications*, 1994;3:27–57.

27 I. Mikic, S. Santini, and R. Jain. Tracking objects in 3D using multiple camera views. In *Proceedings of ACCV*, January 2000: 234–239.

28 S.L. Dockstader and A.M. Tekalp. Multiple camera tracking of interacting and occluded human motion. *Proceedings of the IEEE*, 2001;89(10):1441–1455.

29 A. Mittal and L.S. Davis. M2tracker: a multi-view approach to segmenting and tracking people in a cluttered scene. *International Journal of Computer Vision*, 2003;51(3): 189–203.

30 J.B. Gao and C.J. Harris. Some remarks on Kalman filters for the multisensor fusion. *Information Fusion*, 2002;3(3):191–201.

31 J. Roecker and C. McGillem. Comparison of two-sensor tracking methods based on state vector fusion and measurement fusion. *IEEE Transactions on Aerospace and Electronic Systems*, 1988;24(4):447–449.

32 L. Snidaro, R. Niu, P.K. Varshney, and G.L. Foresti. Sensor fusion for video surveillance. In *Proceedings of the Seventh International Conference on Information Fusion*, Stockholm, Sweden, June 2004, pp. 739–746.

33 D.F. Rogers and J.A. Adams. *Mathematical Elements for Computer Graphics*. McGraw Hill, 1990.

34 T. Wada and T. Matsuyama. Multiobject behavior recognition by event driven selective attention method. *IEEE Transactions on Pattern Analysis and Machine Intelligence*, 2000;22(8):873–887.

35 G. Medioni, I. Cohen, F. Brémond, S. Hongeng, and R. Nevatia. Event detection and analysis from video streams. *IEEE Transactions on Pattern Analysis and Machine Intelligence*, 2001;23(8):873–889.

Epilogue

S. A. Velastin

Before we conclude with some final remarks, it is useful to remind ourselves again of the overall context in which the methods, technologies and proposed solutions presented in the previous chapters take place.

According to data available from the United Nations [1], in 2005, 49.2% of the world's population and 73.3% of that of Europe live in urban environments. Perhaps more interestingly, the same source indicates that in 1950 these figures were 29.1% and 51.2% and predicts that by 2030 these figures would have risen to 60.8% and 79.6%, respectively. Clearly, we are witnessing a major shift in the life style, expectations, economic relations and personal interactions for most of the people of this planet.

Urban life, in conjunction with population increase, is naturally congested. Worldwide, for example, there has been an increase in population density of around 140% over the last 50 years. On the one hand, people are attracted to migrating into the cities in the hope of an improved quality of life and economic prospects. On the other, the size of such migration puts an enormous pressure on town and city managers. The complex interaction of social, environmental, political, legal and economical issues associated with modern life is well beyond the scope of what a book like this could address, but we believe that it is important that those investigating and deploying technical solutions are reasonably informed of the context in which such solutions are likely to operate.

Perhaps the most current issue of public debate is how to strike a balance between an individual's right to privacy and an individual's right to feel and be safe and secure. Measures have been put in place to counteract anti-social and criminal behaviour. Many of these measures are based on what is known as routine activity theory [2]. Part of the strategies for providing safer environments is to establish the presence of a capable guardian, a person or thing that discourages crime from taking place or that takes control when it has occurred. Typical examples include better lighting, regular cleaning, CCTV and police presence.

There has been a major increase in the number of CCTV systems and cameras deployed in urban environments. Estimations of actual numbers vary, but quoting studies conducted by C Norris (Sheffield University), a report by M Frith [3] pointed out that by 2004 the UK had at least 4 285 000 cameras (one for every 14 people), a rate of increase of 400% over three years, about one-fifth of all CCTV cameras worldwide, and that in a city like London an average citizen might be expected to be captured on CCTV cameras up to 300 times a day.

It has become clear that for a CCTV guardian to be capable, it is vital for it to have the ability to detect problems as they arise (or, even better, prevent situations from becoming problems) so as to provide a timely response. People are very good at determining what combination of circumstances represents something unusual or is likely to evolve into an incident. However, given the large number of cameras involved, over large geographically dispersed areas, it is unlikely that such a capability could be provided by human power alone. In fact, asking people to sit in front of TV screens for many largely uneventful hours at a time is a poor way of utilising the advanced cognitive abilities of human beings and of properly empowering them with satisfying and rewarding jobs. What is needed is scalable, robust computer systems that can record evidence and emulate some of the human observation abilities to warn people of threats to safety or security, in a way that is compatible with personal and group liberties and rights.

So, how far are we to achieving this? What has been presented in this book gives an accurate picture of the current state-of-the-art. Thus it can be safely said that we are in some ways close to achieving some key goals but also in some others far from realising practical systems. What follows gives a concluding overview of aspects in which progress has been made and where significant advances are still required.

As discussed in Chapter 2, we have learnt much about the challenges of integrating even simple computer vision systems into environments where the interaction between human beings is crucial and the interfaces between humans and machines are not trivial. A key characteristic of these environments is that the decisions to intervene are rarely taken by a single individual and/or based on visual detection alone. For example, an operator might consider historical data ('is that a hot-spot?'), whether there is another train due in the next 30 s, time of day, availability of ground patrols and so on.

The idea of combining data from multiple sources (data fusion) has been around for some time and effectively applied in specific domains such as mobile robotics. In the context of visual surveillance, in Chapter 6 we saw how people can be tracked as they move from an area covered by one camera to another area covered by another camera, even when the views from such cameras do not overlap. This is a good example of unsupervised measurement of spatial characteristics, that can deal with model and measurement variability (noise) and that can integrate temporal measurements over a wide visual area. Similar motivation (that of covering a wide area and increasing tracking accuracy through multiple cameras) was illustrated in Chapter 8, applied to the tracking of soccer players in a match. It is possible to see that this approach might be adapted to similar enclosed monitoring situations such as to maintain safety in a public swimming pool, school grounds, public children's playgrounds and so on.

A more advanced approach was described in Chapter 9 which highlights a couple of additional important requirements inspired in the way that humans use CCTV. First there is the idea of establishing a hierarchical architecture as a way of dealing with increasing complexity as the number of cameras increases. Secondly, there is a mechanism by which motorised pan-tilt-zoom (PTZ) cameras can work in collaboration with static cameras so that once something of interest has been detected, such as the presence of someone or something, the PTZ can follow and zoom into it so as to get better quality shots. This is exactly what a typical town centre control room operator will do to balance the need to monitor as much space as possible (with a consequent reduction in resolution) and the need to obtain evidence suitable for identification and possible prosecution.

The key process in these proposed methods, and that of the majority of the computer visual surveillance work to date, is that of tracking, i.e. the localisation[1] of each person in the scene followed by correspondence with the data obtained at the previous point in time. The use of multiple views has been shown to be useful in improving the quality of tracking through multiple observations plus the added advantage that results are generally naturally processed in a global (the 'ground plane') rather than image-based set of co-ordinates. The main driver in these efforts is that once tracking can be done reliably, then the interpretation of what is going on in the environment can be based on comparing an object's observed trajectory with what has been pre-established to be normal in that environment. With some understandable degree of optimism, many workers in this field apply terms such as trajectory reasoning or behaviour analysis to what seems to be more closely related to stochastic signal processing.

The reader should be aware that what might seem deceptively simple here, for example, tracking a person as they move in a human-made environment, is at the edge of current advances. For example tracking of people in situations other than from fairly sparse traffic (of up to around 0.1 pedestrians/m^2) is still a major challenge, and to date there are no significant deployments of practical large systems with abilities other than digital storage/retrieval of images, very simple so-called motion detection (the measurement of a significant temporal/spatial change in a video signal) and at best image-based signatures (e.g. colour, blobs), especially associated with MPEG-7 work.

Nevertheless, tracking is a key process to guiding higher-level processes to concentrate on data known to be potentially meaningful. These higher-level processes could allow us in future to extract better clues to human behaviour such as those that might be derived from posture, gesture and the interaction between two or more people or between people and the built environment (e.g. as in graffiti, vandalism or theft). This type of process is still the subject of research.

How information, as distinct from (raw) data, can be combined and tailored to a particular recipient is something that needs further study and development. Such

---

[1] We deliberately use the term 'localisation' (the observation that a person is in a particular place) as opposed to 'identification', normally associated with determining who the person is through some biometric measurement (as in face recognition systems).

efforts are being focused on what is sometimes known as intelligent environments, intelligent spaces, ubiquitous computing and so on, for example, through the ambient intelligence approach [4], a good example of which was given in Chapter 4. The motivation for this type of work is understanding how to fully instrument environments to assist human activity in a way that is transparent to their users. The links with surveillance systems are direct, in particular the need to fully understand human behaviour through a combination of different types of sensors and advanced cognitive ability. This can lead to systems that can interpret what is going on in the ground ('is that person in distress?', 'is the interaction between those two people likely to lead to aggression?'), decide how assistance might be called upon ('how long will it take patrol B to arrive at the scene?', 'let's inform the child's parents') and provide advanced control centres that are aware of how different operators work. How this type of detection and decision-making environment can be realised has been investigated in complex situations such as public transport networks, as exemplified by the PRISMATICA project and described in Chapter 3 and in Reference 5. We note here that in spite of a general recognition that a diversity of sensing modalities is necessary and desirable, most of the work has tended to concentrate on the use of single modalities, especially conventional video.

Even with all the current impressive research work in computer vision systems, their use for surveillance has been very limited. This is illustrated by a recent statement from the UK's Faraday Imaging Partnership [6]:

> Over the past three years, the consistent and widespread message from the end-users of imaging systems has been that this technology is not yet sufficiently robust or generically applicable to reach its full commercial potential. Either systems do not work over a sufficiently wide range of environmental conditions and perturbing factors or they are not applicable in a sufficiently broad range of scenarios without extensive retraining and customisation to each new application

To consider the use of computer vision systems in public spaces for monitoring human activity will stretch these systems to their limit and we hope this will produce progress. For example, a municipality might have typically around 500 cameras that need to be monitored 24 hours a day, under all types of weather conditions and be able to deal with visually complicated and ambiguous patterns of human behaviour. If end-users are to believe that there is potential in these systems, they need to be shown that these systems can operate reliably, with little adjustment, and will not generate a large number of false alarms. Hence, the research community and providers need to make significant efforts to thoroughly evaluate the performance of their proposed solutions with tools such as those outlined in Chapter 3 and efforts such as those of PETS [7], under realistic conditions.

An aspect that is perhaps often overlooked, as mentioned in Chapter 1, is that these systems are increasingly likely to become safety-critical, where guaranteeing a response within a given time is vital. In large heterogeneous systems, communications between modules (e.g. see Chapter 5) needs to be carefully considered. It is important that the computer vision community incorporates systems engineering practices, not simply as a mere implementation issue but at the core of algorithmic designs to make them resilient to data-dependent non-determinism and component failure.

It can be noted that some of the chapters in this book make reference to learning. This is arguably what distinguishes an intelligent system from a mere complex one. What we are asking these systems to do and the range of conditions under which they have to operate are too complex to safely assume that the designers can pre-determine all possible ways in which a system needs to respond. How a system can do this effectively in the contexts discussed here is not yet clear. There is a fine line between a process of measuring (or characterising) something whose model is deter-mined *a priori* and what most people would associate with true learning (a process of discovery, trial and error, rewards, etc.). For example, obtaining a probability distri-bution of average pedestrian speeds in an area (typically to identify cases that depart significantly from the majority) or establishing camera model parameters using what is seen by a camera would not normally be called learning by the general public. On the other hand, the process by which an apprentice[2] control room operator even-tually becomes an expert by observing and emulating people with more experience is what we would normally associate with a true process of learning. As a final remark, many proposed solutions so far do not seem to incorporate an integral measure of the degree of confidence that can be placed on a given result. This can be effective in reducing the practical impact of false alarms and having systems that recognise their own limitations and ask for (human) help when appropriate.

In conclusion, impressive progress is being made towards understanding how to build large inter-connected systems that can monitor human-made spaces (especially human activity), to store meaningful data and to generate alarms upon the detection of something unusual or pre-programmed to need human attention for deciding if intervention is necessary. So, we are not far from achieving distributed surveillance systems. However, to fully earn the title of intelligent distributed surveillance systems, there is significant research to do, to be followed by development and implementations that demonstrate that these systems have a place in improving quality of life in a manner consistent with societal expectations of rights. From a technical point of view, we believe that the key issues that will keep us all busy over the next few years are the following:

- Integrating diverse sensing modalities to the surveillance task (normal range vision, sound, infrared, etc.).
- Sufficient robustness to operate on a continuous basis (through learning, stochastic decision making processes, self-evaluation of trustworthiness, etc.).
- Incorporating contextual knowledge and generating metadata that can later be used for context and/or learning.
- Goal- and event-driven sensing (especially cameras and networks of collaborative cameras).
- Detection of specific behaviour (e.g. aggressiveness), sequences of behaviour characteristic of more complicated patterns ('entered the room and did not seem

---

[2] 'Someone who learns mostly on the job under the direction of a qualified trades person', www.northernopportunities.bc.ca/glossary.html (26th April 2005).

to move and then left without going to the counter'), linking with possible intention.

- Full use of safety-critical systems engineering principles.

We hope that you have enjoyed this book, that it has given a representative description of the state-of-the-art in the field and identified the main current challenges facing those working in this area.

References

1　Population Division of the Department of Economic and Social Affairs of the United Nations Secretariat, World Population Prospects: The 2004 Revision and World Urbanization Prospects: The 2003 Revision, http://esa.un.org/unpp, 22 April 2005.

2　R.V. Clarke and M. Felson (Eds.) Routine activity and rational choice, Vol. 5, *Advances in Criminology Theory*. Transaction Publishers, Inc., New Brunswick, 1993.

3　M. Frith. Big brother Britain, 2004. *The Independent*, 12th January 2004, London.

4　P. Remagnino, G. Foresti, and T. Ellis (Eds.). *Ambient Intelligence A Novel Paradigm*. Springer, 2005.

5　S.A. Velastin, B. Lo, J. Sun, M.A. Vicencio-Silva, and B. Boghossian. PRISMATICA: toward ambient intelligence in public transport environments. *IEEE Systems, Man and Cybernetics (SMC)*, 2005; 35(1):164–182. Special Issue on Ambient Intelligence.

6　Sira Ltd, http://www.imagingfp.org.uk, accessed 26th April 2005.

7　J.M. Ferryman and J. Crowley (Eds.). *Proceedings of the 6th International Workshop on Performance Evaluation of Tracking and Surveillance (PETS'04)*, Prague, Czech Republic, May 2004.

Index